Springer
Berlin
Heidelberg
New York
Hong Kong
London
Milan
Paris
Tokyo

Eugene Fink
Derick Wood

Restricted-Orientation Convexity

With 74 Figures

Authors

Eugene Fink

University of South Florida
Dept. of Computer Science
4204 East Fowler Avenue
Tampa, Florida 33620, USA

Derick Wood

Hong Kong University of Science
and Technology
Dept. of Computer Science
Clear Water Bay, Kowloon
Hong Kong, PR China

Series Editors

Prof. Dr. Wilfried Brauer
Institut für Informatik der TUM
Bolzmannstr. 3, 85748 Garching bei München, Germany
Brauer@informatik.tu-muenchen.de

Prof. Dr. Grzegorz Rozenberg
Leiden Institute of Advanced Computer Science
University of Leiden
Niels Bohrweg 1, 2333 CA Leiden, The Netherlands
rozenber@liacs.nl

Prof. Dr. Arto Salomaa
Turku Centre for Computer Science
Lemminkäisenkatu 14 A, 20520 Turku, Finland
asalomaa@utu.fi

Library of Congress Cataloging-in-Publication Data applied for

Die Deutsche Bibliothek – CIP-Einheitsaufnahme

Bibliographic information published by Die Deutsche Bibliothek
Die Deutsche Bibliothek lists this publication in the Deutsche Nationalbibliografie;
detailed bibliographic data is available in the Internet at <http://dnb.ddb.de>.

ACM Computing Classification (1998): F.2.2

ISBN 3-540-66815-2 Springer-Verlag Berlin Heidelberg New York

Springer-Verlag is a part of Springer Science+Business Media

springeronline.com

Cover Design: KünkelLopka, Heidelberg
Typesetting: Computer to film by author's data
Printed on acid-free paper 45/3142/PS - 5 4 3 2 1 0

To Lena and Mary

Preface

Restricted-orientation convexity, also called $\mathcal{O}$-convexity, is the study of geometric objects whose intersections with lines from some fixed set are connected. This notion generalizes standard convexity and several types of nontraditional convexity. We explore this generalized convexity in multidimensional Euclidean space and identify the properties of standard convex sets that also hold for restricted-orientation convexity.

The purpose of the book is to present the current results on restricted-orientation convexity to the research community and discuss related open problems. The book requires only basic knowledge in geometry; the reader should be familiar with the notion of higher-dimensional Euclidean space and with basic objects in this space, such as lines, balls, and hyperplanes. We use geometric techniques in most proofs, which are accessible to all mathematics and computer-science researchers and graduate students.

$\mathcal{O}$-convexity: We begin with basic properties of $\mathcal{O}$-convex sets, and then introduce $\mathcal{O}$-connected sets, which are a subclass of $\mathcal{O}$-convex sets. We study restricted-orientation analogs of lines, flats and hyperplanes, and characterize $\mathcal{O}$-convex and $\mathcal{O}$-connected sets in terms of their intersections with hyperplanes. We also explore properties of $\mathcal{O}$-connected curves; in particular, we determine when the replacement of a segment of an $\mathcal{O}$-connected curve gives a new $\mathcal{O}$-connected curve, and when the catenation of several curvilinear segments gives an $\mathcal{O}$-connected segment. We use these results to characterize an $\mathcal{O}$-convex set in terms of $\mathcal{O}$-convex segments joining its points, and an $\mathcal{O}$-connected set in terms of $\mathcal{O}$-connected segments.

$\mathcal{O}$-halfspaces: We introduce $\mathcal{O}$-halfspaces, which are a generalization of standard halfspaces, defined as geometric objects whose intersection with every line from some fixed set is empty, a ray or a line. We give basic properties of $\mathcal{O}$-halfspaces and compare them with standard halfspaces; in particular, we show that $\mathcal{O}$-halfspaces may be disconnected and characterize them through their connected components. We also characterize $\mathcal{O}$-halfspaces in terms of $\mathcal{O}$-convexity of their boundaries, and give a condition under which the complement of an $\mathcal{O}$-halfspace is an $\mathcal{O}$-halfspace.

Strong $\mathcal{O}$-convexity: We also introduce the notion of strong $\mathcal{O}$-convexity, which is an alternative generalization of convexity. We describe properties of strongly $\mathcal{O}$-convex flats and halfspaces, and establish the strong $\mathcal{O}$-convexity of the affine hull of a strongly $\mathcal{O}$-convex set. We then show that, for every point in the boundary of a strongly $\mathcal{O}$-convex set, there is a supporting strongly $\mathcal{O}$-convex hyperplane through it. Finally, we characterize strongly $\mathcal{O}$-convex sets in terms of the intersections of strongly $\mathcal{O}$-convex halfspaces.

Acknowledgments: We are grateful to Alex Gurevich for his help with intricate mathematical issues, and to Sven Schuierer for his comments and help in focusing the presentation. The reported work has been supported under grants from the Natural Sciences and Engineering Research Council of Canada, the Information Technology Research Centre of Ontario and the Research Grants Council of Hong Kong. The authors also thank the institutions at which they have done this work, including Carnegie Mellon University, the Hong Kong University of Science and Technology, the University of South Florida, the University of Waterloo and the University of Western Ontario.

October 2003

Eugene Fink
Derick Wood

Contents

1 Introduction

The study of convex sets is a branch of geometry, analysis and linear algebra, which has numerous connections with other areas of mathematics, including topology, number theory and combinatorics [6, 14, 21]. Researchers have explored not only mathematical properties of convex sets, but also related computational problems [5, 13, 34], and applied the resulting algorithms in many practical areas, such as graphics, finite-element analysis, VLSI design and motion planning. They have also studied several types of nontraditional convexity, such as ortho-convexity [28, 30], restricted-orientation convexity [35], NESW convexity [25, 49, 50] and link convexity [2, 52].

The notion of restricted orientations has stemmed from the study of ortho-polygons, which are polygons with edges parallel to the coordinate axes [19]. Researchers have extensively investigated ortho-polygons [1,3,4,11,33,58,59], and used them in geometric models based on vertical and horizontal lines, such as VLSI wiring and architectural floor plans. They have also studied ortho-convex sets, which are sets whose intersection with every vertical and every horizontal line is connected [28, 30, 32, 39, 42].

Güting introduced restricted orientations as a generalization of ortho-polygons [16]; he explored computational properties of polygons whose edges were parallel to the elements of some fixed set of lines [16–18]. Widmayer, Wu, Schlag and Wong also studied computational problems related to restricted orientations [55–57]. Nilsson, Ottmann, Schuierer and Icking reviewed and extended the earlier results in restricted-orientation geometry [31].

Rawlins and Wood used restricted orientations to define the notion of $\mathcal{O}$-convexity, which generalized standard convexity and ortho-convexity [35, 37–41]. Schuierer continued their exploration and presented an extensive study of geometric and computational properties of $\mathcal{O}$-convex sets [43]. Rawlins introduced an alternative generalization of convexity based on restricted orientations, called strong $\mathcal{O}$-convexity [35]. We considered computational problems in strong $\mathcal{O}$-convexity and developed a suite of related algorithms [10].

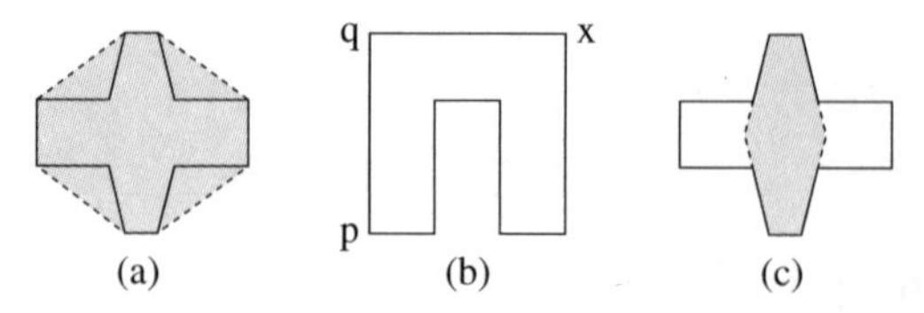

Fig. 1.1. Standard convex hull **(a)** and standard kernels **(b,c)**

Although researchers have extensively studied nontraditional convexity in the plane, they have not extended it to higher dimensions. The purpose of our work is to develop a theory of restricted-orientation convexity in multidimensional space [7–9].

We begin with a review of standard convexity, and define the related notions of convex hulls and kernels (Sect. 1.1). We also review ortho-convexity and strong ortho-convexity, which are special cases of restricted-orientation convexity (Sects. 1.2 and 1.3), and define a topological generalization of convex sets (Sect. 1.4). We then outline the organization of the book and dependencies between its chapters (Sect. 1.5).

1.1 Standard Convexity

We review basic properties of convex sets in the plane; a much more extended review is available in several texts on convexity, including *Convex Polytopes* by Grünbaum, Klee, Perles and Shephard [15], *Geometry and Convexity* by Kelly and Weiss [20], *Convex Sets* by Valentine [51] and *Convexity* by Webster [54].

We define convex sets through their intersections with lines; specifically, a set is **convex** if its intersection with every line is connected.

Proposition 1.1 (Properties of standard convex sets).

1. *The intersection of convex sets is a convex set.*
2. *Every convex set is simply connected.*
3. *A closed set is convex if and only if it is either the entire plane or the intersection of halfplanes.*

The **convex hull** of a geometric object is the intersection of all convex sets that contain the object; for example, the shaded region in Fig. 1.1a is the convex hull of the polygon shown by solid lines.

Proposition 1.2 (Properties of standard convex hulls).

1. *The convex hull of a geometric object is the minimal convex set that contains the object.*

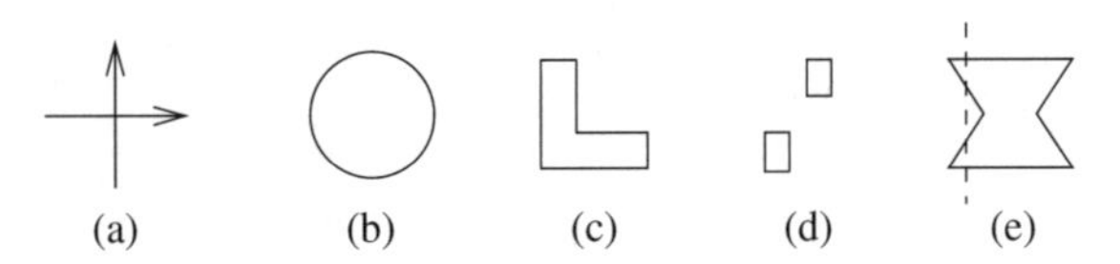

Fig. 1.2. Ortho-convexity

2. *An object is convex if and only if it is identical to its convex hull.*

Two points of a geometric object are **visible** to each other if the line segment joining them is wholly in the object; for example, the points p and q of the polygon in Fig. 1.1b are visible to each other, whereas p and x are not. Note that an object is convex if and only if every two of its points are visible to each other. The **kernel** of a geometric object is the set of points that are visible from all points of the object; for example, the kernel of the polygon in Fig. 1.1b is empty, whereas the kernel of the polygon in Fig. 1.1c is the nonempty shaded region.

Proposition 1.3 (Properties of standard kernels).

1. *The kernel of any geometric object is convex.*
2. *An object is convex if and only if it is identical to its kernel.*

1.2 Ortho-Convexity

We now consider ortho-convexity, which is weaker than standard convexity. A set is **ortho-convex** if its intersection with every vertical line and every horizontal line is connected. For example, the sets in Fig. 1.2b–d are ortho-convex, whereas the set in Fig. 1.2e is not ortho-convex, since its intersection with the dashed vertical line is disconnected. Note that ortho-convex sets may be disconnected; for instance, the set in Fig. 1.2d consists of two components.

Proposition 1.4 (Properties of ortho-convex sets).

1. *The intersection of ortho-convex sets is an ortho-convex set.*
2. *Every standard convex set is ortho-convex.*
3. *A disconnected set is ortho-convex if and only if every connected component of the set is ortho-convex and no vertical or horizontal line intersects two components.*
4. *Every connected ortho-convex set is simply connected.*

The **ortho-hull** of a geometric object is the intersection of all ortho-convex sets that contain the object; we give four examples of ortho-hulls in Fig. 1.3.

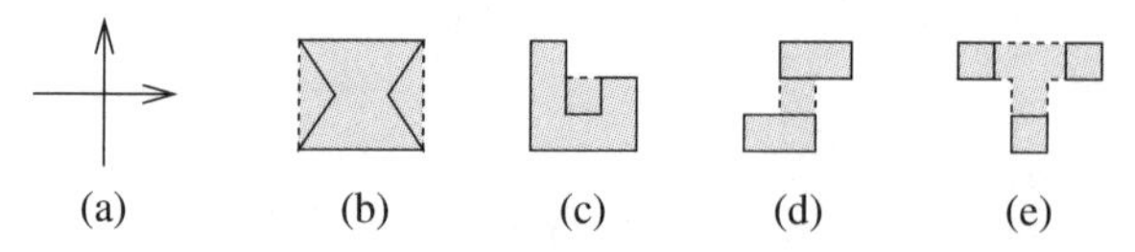

Fig. 1.3. Ortho-hulls of two connected sets **(b,c)** and two disconnected sets **(d,e)**

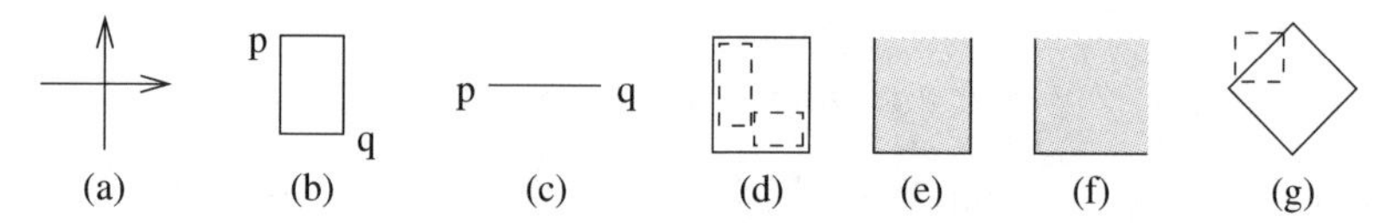

Fig. 1.4. Strong ortho-convexity

Proposition 1.5 (Properties of ortho-hulls).

1. *The ortho-hull of a geometric object is the minimal ortho-convex set that contains the object.*
2. *An object is ortho-convex if and only if it is identical to its ortho-hull.*
3. *The ortho-hull of an object is a subset of the standard convex hull of the object.*

1.3 Strong Ortho-Convexity

We next review a different type of nontraditional convexity, which is also defined through vertical and horizontal lines, and consider the related notions of ortho-rectangles and ortho-blocks. An **ortho-rectangle** is a rectangle whose sides are parallel to the coordinate axes. An **ortho-block** of two points p and q is the minimal ortho-rectangle that contains them; note that p and q are opposite vertices of this rectangle, as shown in Fig. 1.4b. In particular, if p and q are on the same vertical or horizontal line, their ortho-block is the line segment joining them, as shown in Fig. 1.4c.

A set is **strongly ortho-convex** if, for every two of its points, their ortho-block is wholly in the set. For example, the rectangle in Fig. 1.4d is strongly ortho-convex; two ortho-blocks contained in this rectangle are shown by dashed lines. As another example, the unbounded sets in Fig. 1.4e,f are also strongly ortho-convex. On the other hand, the square in Fig. 1.4g is not strongly ortho-convex, because the dashed ortho-block is not in this square.

Proposition 1.6 (Properties of strongly ortho-convex sets).

1. *The intersection of strongly ortho-convex sets is a strongly ortho-convex set.*
2. *Every strongly ortho-convex set is standard convex.*
3. *Every strongly ortho-convex set is simply connected.*

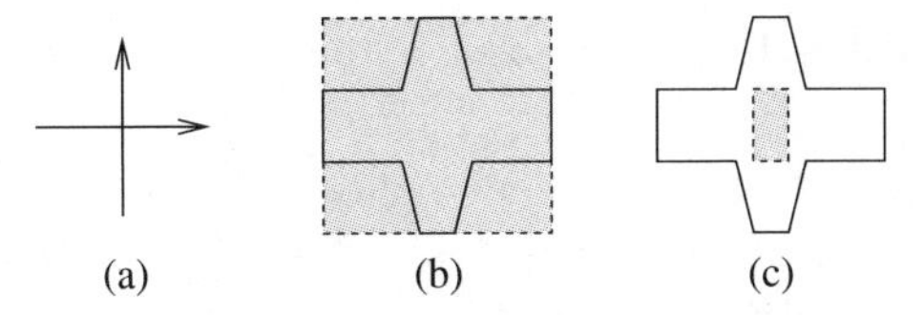

Fig. 1.5. Strong ortho-hull **(b)** and strong ortho-kernel **(c)**

4. *A halfplane is strongly ortho-convex if and only if its boundary line is vertical or horizontal.*
5. *A closed set is strongly ortho-convex if and only if it is either the entire plane or the intersection of strongly ortho-convex halfplanes.*
6. *A closed bounded set is strongly ortho-convex if and only if it is an ortho-rectangle.*

The **strong ortho-hull** of a geometric object is the intersection of all strongly ortho-convex sets that contain the object, as illustrated in Fig. 1.5b.

Proposition 1.7 (Properties of strong ortho-hulls).

1. *The strong ortho-hull of a geometric object is the minimal strongly ortho-convex set that contains the object.*
2. *An object is strongly ortho-convex if and only if it is identical to its strong ortho-hull.*
3. *The standard convex hull of an object is a subset of the strong ortho-hull of the object.*

We can define strong ortho-visibility in terms of ortho-blocks; that is, two points of a geometric object are strongly ortho-visible to each other if their ortho-block is wholly in the object. Note that an object is strongly ortho-convex if and only if every two of its points are strongly ortho-visible to each other. The **strong ortho-kernel** of a geometric object is the set of points that are strongly ortho-visible from every point of the object; we give an example of a strong ortho-kernel in Fig. 1.5c.

Proposition 1.8 (Properties of strong ortho-kernels).

1. *The strong ortho-kernel of any geometric object is strongly ortho-convex.*
2. *An object is strongly ortho-convex if and only if it is identical to its strong ortho-kernel.*
3. *The strong ortho-kernel of an object is a subset of the standard kernel of the object.*

Table 1.1. Comparison of different convexities

Intersection	
Standard convexity:	The intersection of convex sets is a convex set.
Ortho-convexity:	The intersection of ortho-convex sets is an ortho-convex set.
Strong ortho-convexity:	The intersection of strongly ortho-convex sets is a strongly ortho-convex set.
Line intersection	
Standard convexity:	A set is convex if and only if its intersection with every line is connected.
Ortho-convexity:	A set is ortho-convex if and only if its intersection with every vertical line and every horizontal line is connected.
Connectedness	
Standard convexity:	Every convex set is simply connected.
Ortho-convexity:	Every connected ortho-convex set is simply connected.
Strong ortho-convexity:	Every strongly ortho-convex set is simply connected.
Visibility	
Standard convexity:	A set is convex if and only if, for every two of its points, the line segment joining them is wholly in the set.
Strong ortho-convexity:	A set is strongly ortho-convex if and only if, for every two of its points, their ortho-block is wholly in the set.
Kernel convexity	
Standard convexity:	The standard kernel of any set is convex.
Ortho-convexity:	The ortho-kernel of any set is ortho-convex.
Strong ortho-convexity:	The strong ortho-kernel of any set is strongly ortho-convex.
Halfspace intersection	
Standard convexity:	A closed set is convex if and only if it is either the entire plane or the intersection of halfplanes.
Strong ortho-convexity:	A closed set is strongly ortho-convex if and only if it is either the entire plane or the intersection of strongly ortho-convex halfplanes.

1.4 Convexity Spaces

The properties of ortho-convexity and strong ortho-convexity are similar to those of standard convexity, as shown in Table 1.1. The basic results for other types of nontraditional convexity are also analogous to those for standard convexity. This similarity has led to the notion of convexity spaces [24], which is a topological generalization of convex sets; a review of the related results is available in the book by van de Vel [53].

A **convexity space** is defined by two sets, X and $\mathcal{C}$, where X is an arbitrary set, and $\mathcal{C}$ is a collection of subsets of X that satisfies two conditions:

1. The empty set and the entire set X are elements of $\mathcal{C}$.
2. For every subset C of $\mathcal{C}$, the intersection $\cap C$ of its elements is in $\mathcal{C}$.

Informally, X is an analog of the plane in standard convexity, and the elements of $\mathcal{C}$ are analogs of convex sets, which are called $\mathcal{C}$-convex sets. The two conditions generalize the observation that the empty set and the entire plane are convex, and the intersection of convex sets is a convex set.

The related definition of a hull is the same as in standard convexity; that is, for every subset Y of X, the **$\mathcal{C}$-hull** of Y is the intersection of all $\mathcal{C}$-convex sets that contain Y.

Proposition 1.9 (Properties of $\mathcal{C}$-hulls).

1. *The $\mathcal{C}$-hull of a subset Y of X is the minimal $\mathcal{C}$-convex set that contains Y.*
2. *A subset Y of X is $\mathcal{C}$-convex if and only if it is identical to its $\mathcal{C}$-hull.*

Schuierer, Rawlins and Wood defined visibility in convexity spaces and studied its properties [35,36,43,44,48,60]. Two elements p and q of a subset Y of X are visible to each other if the $\mathcal{C}$-hull of the two-element set $\{p, q\}$ is wholly in Y. Note that the hull of two points in standard convexity is the line segment joining them, and the strong ortho-hull of two points is their ortho-block, which means that visibility in convexity spaces generalizes standard visibility and strong ortho-visibility.

1.5 Book Outline

We present the results of exploring two notions of nontraditional convexity in multidimensional space, called $\mathcal{O}$-convexity and strong $\mathcal{O}$-convexity, which also satisfy the general conditions of convexity spaces. These two notions generalize standard convexity, ortho-convexity, and strong ortho-convexity. In Fig. 1.6, we summarize the organization of the book.

We first describe the properties of $\mathcal{O}$-convexity and strong $\mathcal{O}$-convexity in two dimensions (Chap. 2) and consider related computational problems (Chap. 3). We then generalize $\mathcal{O}$-convexity to higher dimensions, and show that the properties of the resulting generalization are much richer than those in two dimensions (Chap. 4). We also consider $\mathcal{O}$-convexity analogs of halfspaces and study their relationship to $\mathcal{O}$-convex sets (Chap. 5). Finally, we extend strong $\mathcal{O}$-convexity to higher dimensions, describe the main properties of strongly $\mathcal{O}$-convex sets, and give additional properties of strongly $\mathcal{O}$-convex flats and halfspaces (Chap. 6). We conclude with a summary of results and related open problems (Chap. 7).

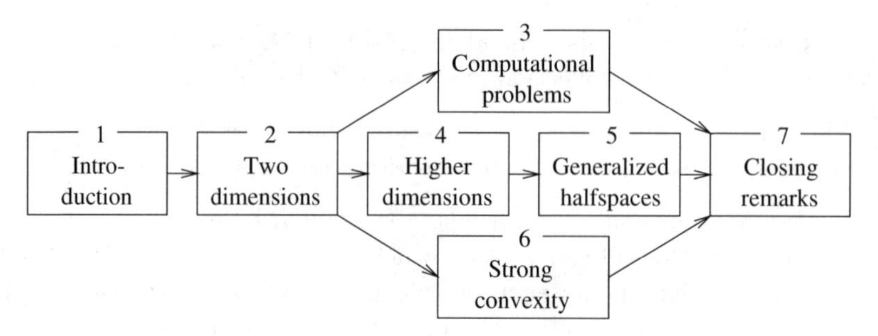

Fig. 1.6. Dependencies among chapters

2

Two Dimensions

We begin with two planar generalizations of convexity, called $\mathcal{O}$-convexity and strong $\mathcal{O}$-convexity. We first define $\mathcal{O}$-convex sets and present their basic properties (Sect. 2.1), then introduce a restricted-orientation analog of halfplanes (Sect. 2.2), and finally describe strong $\mathcal{O}$-convexity (Sect. 2.3).

2.1 $\mathcal{O}$-Convex Sets

Rawlins introduced the notion of planar $\mathcal{O}$-convexity in 1987, as a generalization of ortho-convexity [32] and standard convexity. He defined $\mathcal{O}$-convex sets in terms of their intersections with lines by analogy with one of the definitions of standard convex sets. Rawlins, Schuierer and Wood explored properties of $\mathcal{O}$-convex sets in two dimensions and demonstrated their similarity to standard convex sets [39, 41, 43].

Recall that convex sets can be described through their intersections with lines; specifically, a set is convex if its intersection with every line is connected. We define $\mathcal{O}$-convex sets through their intersections with lines in a given set rather than with all lines. To define such a restricted collection of lines, we first introduce the notion of an **orientation set** $\mathcal{O}$, which is a (possibly infinite) set of lines through some fixed point o; we give an example of a finite orientation set in Fig. 2.1a. A line that is a translate of an element of $\mathcal{O}$ is called an **$\mathcal{O}$-line;** for example, the dashed lines in Fig. 2.1b–e are $\mathcal{O}$-lines. We use the collection of all translates of all lines in a given $\mathcal{O}$ to define $\mathcal{O}$-convex sets.

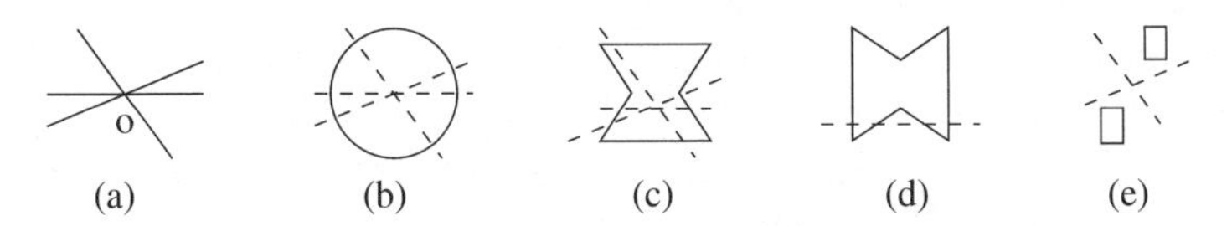

Fig. 2.1. Planar $\mathcal{O}$-convexity

Definition 2.1 ($\mathcal{O}$-convexity). *A set is* **$\mathcal{O}$-convex** *if its intersection with every $\mathcal{O}$-line is connected.*

For the orientation set in Fig. 2.1a, the objects in Fig. 2.1b,c are $\mathcal{O}$-convex; some $\mathcal{O}$-lines intersecting them are shown by dashed lines. On the other hand, the object in Fig. 2.1d is not $\mathcal{O}$-convex, since its intersection with the dashed $\mathcal{O}$-line is disconnected. Note that the object in Fig. 2.1d is a rotation of that in Fig. 2.1c, which shows that rotations may not preserve $\mathcal{O}$-convexity. Unlike standard convex sets, $\mathcal{O}$-convex sets may be disconnected; for example, the two rectangles in Fig. 2.1e form a disconnected $\mathcal{O}$-convex set. We now give some basic properties of planar $\mathcal{O}$-convex sets [41].

Lemma 2.1.

1. *Every translate of an $\mathcal{O}$-convex set is $\mathcal{O}$-convex.*
2. *If C is a collection of $\mathcal{O}$-convex sets, then the intersection $\bigcap C$ of these sets is also an $\mathcal{O}$-convex set.*
3. *Every standard convex set is $\mathcal{O}$-convex.*
4. *If $\mathcal{O}_1 \subseteq \mathcal{O}_2$, then every $\mathcal{O}_2$-convex set is $\mathcal{O}_1$-convex.*
5. *A disconnected set is $\mathcal{O}$-convex if and only if every connected component of the set is $\mathcal{O}$-convex and no $\mathcal{O}$-line intersects two components.*
6. *If $\mathcal{O}$ is nonempty, then every connected $\mathcal{O}$-convex set is simply connected.*

Proof.

(1) By definition, every translate of an $\mathcal{O}$-line is an $\mathcal{O}$-line. Therefore, if the intersection of a set with every $\mathcal{O}$-line is connected, then the same holds for every translate of the set.

(2) If C is a collection of $\mathcal{O}$-convex sets, then, for every $\mathcal{O}$-line l, the intersection of each element of C with l is connected; hence, the intersection of $\bigcap C$ with l is also connected. We conclude that the intersection of $\bigcap C$ with every $\mathcal{O}$-line is connected, which implies that $\bigcap C$ is $\mathcal{O}$-convex.

(3) The intersection of a convex set with every line is connected. In particular, its intersection with every $\mathcal{O}$-line is connected, which implies that it is $\mathcal{O}$-convex.

(4) If $\mathcal{O}_1 \subseteq \mathcal{O}_2$, then every $\mathcal{O}_1$-line is an $\mathcal{O}_2$-line. The intersection of an $\mathcal{O}_2$-convex set with every $\mathcal{O}_2$-line is connected, which implies that its intersection with every $\mathcal{O}_1$-line is connected; thus, it is $\mathcal{O}_1$-convex.

(5) If a set P is the union of disjoint $\mathcal{O}$-convex components and no $\mathcal{O}$-line intersects two components, then the intersection of P with every $\mathcal{O}$-line is connected; therefore, P is $\mathcal{O}$-convex. If one of P's components is not $\mathcal{O}$-convex, then the intersection of this component with some $\mathcal{O}$-line is disconnected. The intersection of P with this $\mathcal{O}$-line is also disconnected; hence, P is not $\mathcal{O}$-convex. Finally, if some $\mathcal{O}$-line intersects two or more components, then the intersection of P with this $\mathcal{O}$-line is disconnected; therefore, we again conclude that P is not $\mathcal{O}$-convex.

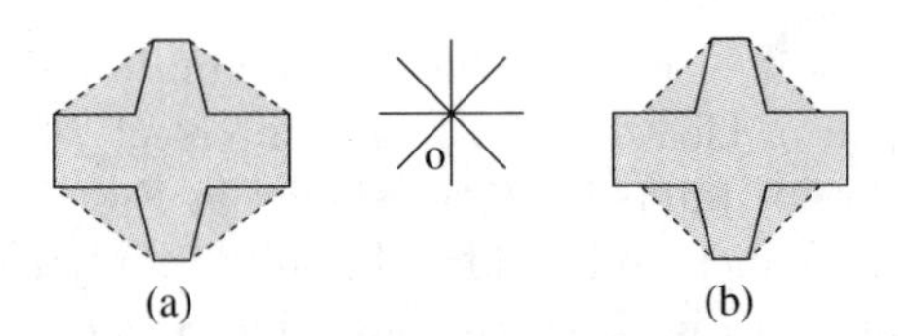

Fig. 2.2. Standard convex hull **(a)** and $\mathcal{O}$-hull **(b)**

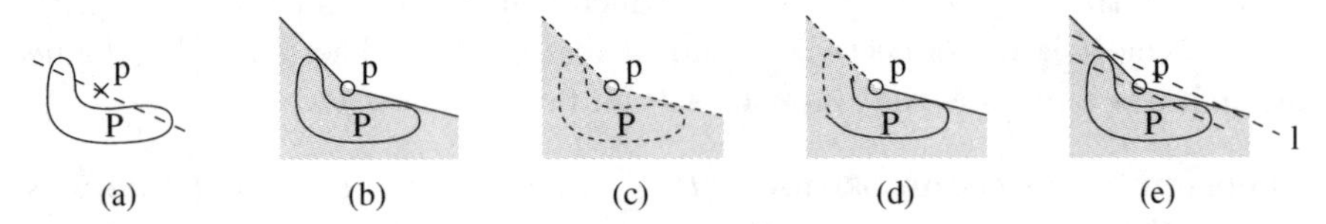

Fig. 2.3. Proof of Theorem 2.3

(6) If a set P is connected but not simply connected, then P has a hole, and there is an $\mathcal{O}$-line that cuts through the hole. The intersection of P with this $\mathcal{O}$-line is disconnected; thus, P is not $\mathcal{O}$-convex. □

We now introduce the notion of an $\mathcal{O}$-hull. Recall that the standard convex hull of a geometric object is the intersection of all convex sets containing the object. Similarly, the **$\mathcal{O}$-hull** of an object is the intersection of all $\mathcal{O}$-convex sets that contain the object. We show a standard convex hull in Fig. 2.2a and an $\mathcal{O}$-hull in Fig. 2.2b. We list basic properties of $\mathcal{O}$-hulls, which immediately follow from the definition.

Lemma 2.2.

1. *The $\mathcal{O}$-hull of a geometric object contains the object.*
2. *A geometric object is $\mathcal{O}$-convex if and only if it is identical to its $\mathcal{O}$-hull.*
3. *The $\mathcal{O}$-hull of a geometric object is a subset of the standard convex hull of the object.*
4. *If $\mathcal{O}_1 \subseteq \mathcal{O}_2$, then the $\mathcal{O}_1$-hull of a geometric object is a subset of the $\mathcal{O}_2$-hull of the object.*

We now establish separation and decomposition properties of $\mathcal{O}$-hulls [40,41].

Theorem 2.3 (Separation). *Suppose that P is a connected set and p is a point outside of P. Then, $p \in \mathcal{O}\text{-hull}(P)$ if and only if there is an $\mathcal{O}$-line through p that intersects P on both sides of p.*

Proof. If an $\mathcal{O}$-line through p intersects P on both sides of p, as shown in Fig. 2.3a, then every $\mathcal{O}$-convex set that contains P also includes p, which implies that $p \in \mathcal{O}\text{-hull}(P)$.

To show the converse, suppose that $p \in \mathcal{O}\text{-hull}(P)$. We draw the two rays from p that support P, as shown in Fig. 2.3b, and consider the shaded angle, which does not include its vertex p. The exact definition of this angle depends on whether P is open or closed. If both rays intersect P, they belong to the angle. If the rays do not intersect P, as shown in Fig. 2.3c, the angle is an open set that does not include its sides. Finally, if only one of the two rays intersects P, as shown in Fig. 2.3d, the angle includes one of its sides.

In all cases, the angle contains P and does not include p. Since $p \in \mathcal{O}\text{-hull}(P)$, the angle is not $\mathcal{O}$-convex, which means that its intersection with some $\mathcal{O}$-line l is disconnected, as shown in Fig. 2.3e. The parallel-to-l line through p is an $\mathcal{O}$-line that intersects P on both sides of p. □

Theorem 2.4 (Decomposition). *If $\mathcal{O}_1$ and $\mathcal{O}_2$ are two orientation sets through the same point o, then, for every connected set P,*

$$(\mathcal{O}_1 \cup \mathcal{O}_2)\text{-hull}(P) = \mathcal{O}_1\text{-hull}(\mathcal{O}_2\text{-hull}(P)) = \mathcal{O}_1\text{-hull}(P) \cup \mathcal{O}_2\text{-hull}(P).$$

Proof. We readily conclude from Lemma 2.2 that

$$\mathcal{O}_1\text{-hull}(P) \cup \mathcal{O}_2\text{-hull}(P) \subseteq \mathcal{O}_1\text{-hull}(\mathcal{O}_2\text{-hull}(P)) \subseteq (\mathcal{O}_1 \cup \mathcal{O}_2)\text{-hull}(P).$$

We now show that, if a point p is in $(\mathcal{O}_1 \cup \mathcal{O}_2)\text{-hull}(P)$, then it is also in $\mathcal{O}_1\text{-hull}(P) \cup \mathcal{O}_2\text{-hull}(P)$. By Theorem 2.3, if p is in $(\mathcal{O}_1 \cup \mathcal{O}_2)\text{-hull}(P)$, then some $\mathcal{O}_1$-line or $\mathcal{O}_2$-line through p intersects P on both sides of p, which implies that p is in $\mathcal{O}_1\text{-hull}(P)$ or in $\mathcal{O}_2\text{-hull}(P)$. □

2.2 $\mathcal{O}$-Halfplanes

Standard halfplanes can also be characterized through their intersections with lines; specifically, a closed set is a halfplane only if its intersection with every line is empty, a ray or a line. We use this observation to define an $\mathcal{O}$-convexity analog of halfplanes.

Definition 2.2 ($\mathcal{O}$-halfplanes). *An* **$\mathcal{O}$-halfplane** *is a closed set whose intersection with every $\mathcal{O}$-line is empty, a ray or a line.*

Note that the *empty set* and the *whole plane* are considered $\mathcal{O}$-halfplanes, which simplifies some definitions and results. For example, the objects in Fig. 2.4b–f are $\mathcal{O}$-halfplanes for the orientation set in Fig. 2.4a; note that the $\mathcal{O}$-halfplane in Fig. 2.4f is disconnected. As another example, the objects in Fig. 2.4h,i are $\mathcal{O}$-halfplanes for the orientation set in Fig. 2.4g.

This notion of $\mathcal{O}$-halfplanes is different from the $\mathcal{O}$-convexity analogs of halfplanes in the work of Rawlins, who defined an **$\mathcal{O}$-stairhalfplane** as a region of the plane bounded by an $\mathcal{O}$-convex curve [35]. $\mathcal{O}$-stairhalfplanes are a proper subclass of $\mathcal{O}$-halfplanes as follows from Lemma 5.8 (page 60).

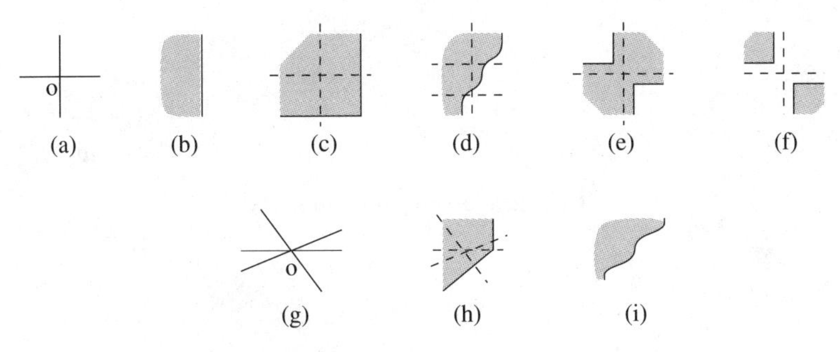

Fig. 2.4. $\mathcal{O}$-halfplanes

Lemma 2.5.

1. *Every translate of an $\mathcal{O}$-halfplane is an $\mathcal{O}$-halfplane.*
2. *Every standard closed halfplane is an $\mathcal{O}$-halfplane.*
3. *Every $\mathcal{O}$-halfplane is $\mathcal{O}$-convex.*
4. *A disconnected set is an $\mathcal{O}$-halfplane if and only if each of its connected components is an $\mathcal{O}$-halfplane and no $\mathcal{O}$-line intersects two components.*

Proof.

(1) If the intersection of a set with every $\mathcal{O}$-line is empty, a ray or a line, then the same holds for every translate of the set.

(2) The intersection of a standard halfplane with every line is empty, a ray or a line; hence, it is an $\mathcal{O}$-halfplane.

(3) The intersection of an $\mathcal{O}$-halfplane with every $\mathcal{O}$-line is connected; therefore, every $\mathcal{O}$-halfplane is $\mathcal{O}$-convex.

(4) If P is the union of disjoint $\mathcal{O}$-halfplanes and no $\mathcal{O}$-line intersects two of them, then the intersection of P with every $\mathcal{O}$-line is empty, a ray or a line; hence, P is an $\mathcal{O}$-halfplane. If one of P's components is not an $\mathcal{O}$-halfplane, the intersection of this component with some $\mathcal{O}$-line is not empty, not a ray and not a line. Then, the intersection of P with this line is not empty, not a ray and not a line; thus, P is not an $\mathcal{O}$-halfplane. Finally, if some $\mathcal{O}$-line intersects two components, then its intersection with P is disconnected, which implies that P is not an $\mathcal{O}$-halfplane. □

We characterize closed $\mathcal{O}$-convex sets in terms of the intersections of $\mathcal{O}$-halfplanes [41].

Lemma 2.6. *A closed connected set is $\mathcal{O}$-convex if and only if it is the intersection of $\mathcal{O}$-halfplanes.*

Proof. Suppose that a set P is the intersection of $\mathcal{O}$-halfplanes. Since every $\mathcal{O}$-halfplane is $\mathcal{O}$-convex by Lemma 2.5, their intersection P is also $\mathcal{O}$-convex by Lemma 2.1.

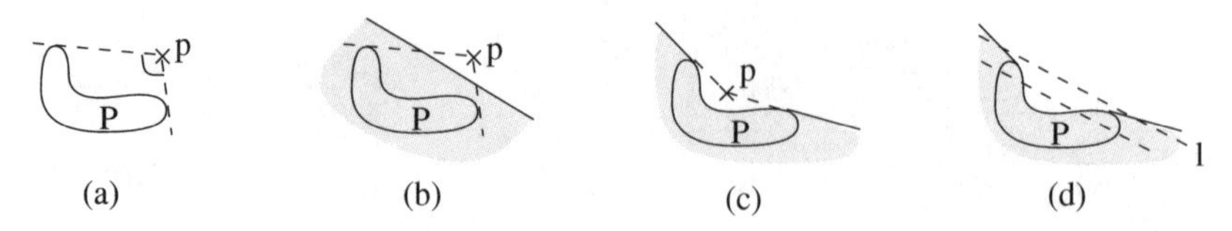

Fig. 2.5. Proof of Lemma 2.6

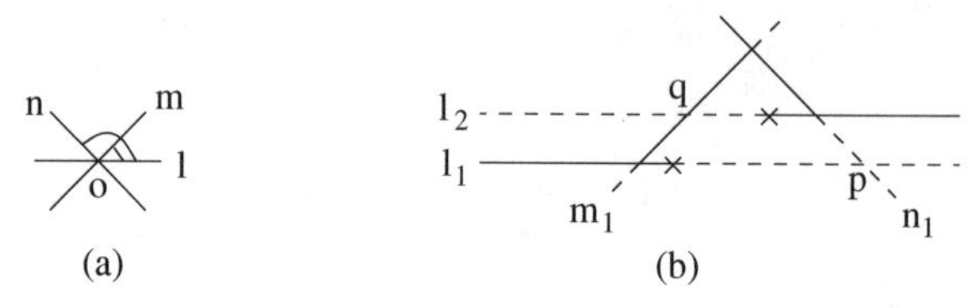

Fig. 2.6. Proof of Lemma 2.7

Now suppose, conversely, that P is $\mathcal{O}$-convex. We show that P is the intersection of $\mathcal{O}$-halfplanes by demonstrating that, for every point p outside of P, some $\mathcal{O}$-halfplane contains P and does not contain p.

We draw the two lines through p that support P, as shown in Fig. 2.5a. If the marked angle between these lines is less than π, there is a standard halfplane that contains P and does not contain p, as shown in Fig. 2.5b.

If the marked angle is at least π, we consider the set Q shown by shading in Fig. 2.5c. The boundary of Q consists of the segment of P's boundary between the supporting lines and the parts of the supporting lines that extend this segment. We show that Q is an $\mathcal{O}$-halfplane.

If the intersection of Q with some $\mathcal{O}$-line l is disconnected, then there is an $\mathcal{O}$-line parallel to l whose intersection with P is disconnected, as shown in Fig. 2.5d, contradicting the assumption that P is $\mathcal{O}$-convex. Furthermore, there is no line whose intersection with Q is a point or segment. Therefore, the intersection of Q with every $\mathcal{O}$-line is empty, a ray or a line. □

If $\mathcal{O}$ contains at least *three distinct lines*, then $\mathcal{O}$-halfplanes have additional basic properties. To derive these properties, we use the notion of the **direction** of a ray; specifically, two rays have the same direction if they are translates of each other.

Lemma 2.7. *Suppose that the orientation set $\mathcal{O}$ contains at least three distinct lines. If the intersection of an $\mathcal{O}$-halfplane with two parallel $\mathcal{O}$-lines forms two rays, these rays have the same direction, rather than opposite directions.*

Proof. Suppose that the intersection of an $\mathcal{O}$-halfplane P with parallel $\mathcal{O}$-lines l_1 and l_2 gives rays of opposite directions; we show these two rays by solid lines in Fig. 2.6b. For convenience, we assume that l_1 is below l_2 and the lower ray's direction is to the left.

Fig. 2.7. Directed $\mathcal{O}$-halfplane **(a)** and nondirected $\mathcal{O}$-halfplane **(b)**

Let l be the element of $\mathcal{O}$ parallel to l_1 and l_2, and let m and n be two other elements of $\mathcal{O}$, as shown in Fig. 2.6a. We assume that the marked angle between l and m is smaller than the marked angle between l and n.

We choose a point $p \in l_1$ and draw a line n_1 through p parallel to n. We select this point p in such a way that p is not in P and n_1 intersects the upper ray, as shown in Fig. 2.6b. Since n_1 is an $\mathcal{O}$-line, its intersection with P must be empty, a ray or a line; hence, the part of n_1 above l_2 (shown by a solid line) is in P.

We next choose a point $q \in l_2$ and draw a line m_1 through q parallel to m. We pick q is such a way that q is not in P and m_1 intersects the lower ray. Note that m_1 intersects n_1 *above* l_2, which implies that m_1 intersects the part of n_1 contained in P. Since m_1 is an $\mathcal{O}$-line, its intersection with P is connected; therefore, the segment of m_1 between l_1 and n_1 is in P, contradicting the assumption that q is not in P. □

We illustrate the directed-ray property of $\mathcal{O}$-halfplanes in Fig. 2.7a, where the intersection of an $\mathcal{O}$-halfplane with several parallel $\mathcal{O}$-lines is shown by dashed rays. The $\mathcal{O}$-halfplanes that satisfy this property are called **directed $\mathcal{O}$-halfplanes.** If $\mathcal{O}$ contains two lines, an $\mathcal{O}$-halfplane may not be directed; for example, the $\mathcal{O}$-halfplane in Fig. 2.7b is not directed, since the dashed $\mathcal{O}$-rays have opposite directions.

Lemma 2.8. *Suppose $\mathcal{O}$ contains at least two distinct lines. Then:*

1. *Every $\mathcal{O}$-halfplane is either connected or consists of two components.*
2. *Every directed $\mathcal{O}$-halfplane is connected.*
3. *The boundary of every directed $\mathcal{O}$-halfplane is connected and $\mathcal{O}$-convex.*

Proof.

(1) We prove that every $\mathcal{O}$-halfplane P has at most two components by showing that, for every three points $p, q, a \in P$, two of them are in the same components.

Let l and m be two elements of $\mathcal{O}$, and suppose for convenience that l is horizontal, as shown in Fig. 2.8a. Since P is an $\mathcal{O}$-halfplane, one of the two horizontal rays with endpoint p is contained in P; we show this ray in Fig. 2.8b. Similarly, we can choose a horizontal ray with endpoint q and a horizontal ray with endpoint a contained in P.

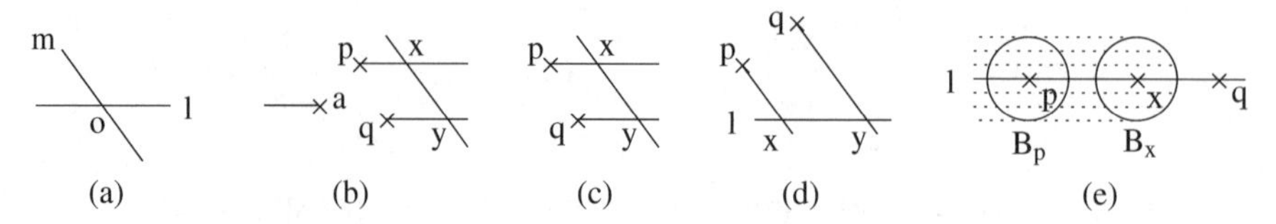

Fig. 2.8. Proof of Lemma 2.8

We select two of these three rays that have the same direction; without loss of generality, assume that the endpoints of the selected rays are p and q. We choose a parallel-to-m line that intersects these two rays, and denote the respective intersection points by x and y, as shown in Fig. 2.8b. The polygonal line (p, x, y, q) is wholly in P, which implies that p and q are in the same connected component.

(2) We show that every two points p and q of a directed $\mathcal{O}$-halfplane P can be connected by a polygonal line in P. We pick two parallel $\mathcal{O}$-rays, with endpoints p and q, that are contained in P and have the same direction, and consider an $\mathcal{O}$-line that intersects these rays. We illustrate this construction in Fig. 2.8c, where the respective intersection points are denoted by x and y. The polygonal line (p, x, y, q) is a path from p to q within P.

(3) Suppose that the boundary of a directed $\mathcal{O}$-halfplane P is not connected. Since P is connected, the complement of P is disconnected and we can choose points p and q in different connected components of P's complement. Next, we pick two parallel $\mathcal{O}$-rays, with endpoints p and q, that do not intersect P and have the same direction, as shown in Fig. 2.8d. Finally, we select an $\mathcal{O}$-line l that intersects these two rays, and denote the respective intersection points by x and y. The segment of l between x and y does not intersect P, because, if some point z of this segment were in P, then one of the two contained-in-l rays with endpoint z would be in P, contradicting the assumption that x and y are not in P. Therefore, the polygonal line (p, x, y, q) is wholly in P's complement, contradicting the assumption that p and q are in different components of P's complement.

Now suppose that the boundary of P is not $\mathcal{O}$-convex. Then, the intersection of some $\mathcal{O}$-line l with P's boundary is disconnected, and we can select points $p, q \in l$ that are in the boundary and a point $x \in l$ between them that is not in the boundary; we assume that p is to the left of x, as shown in Fig. 2.8e. Since the intersection of P with l is connected, x is in the interior of P, and we can choose a circle $B_x \subseteq P$ centered at x. Either all left-directed or all right-directed rays with endpoints in B_x are contained in P; we assume that the left-directed rays are in P. Then, some circle B_p centered at p is wholly in P; therefore, p is in P's interior, which yields a contradiction. □

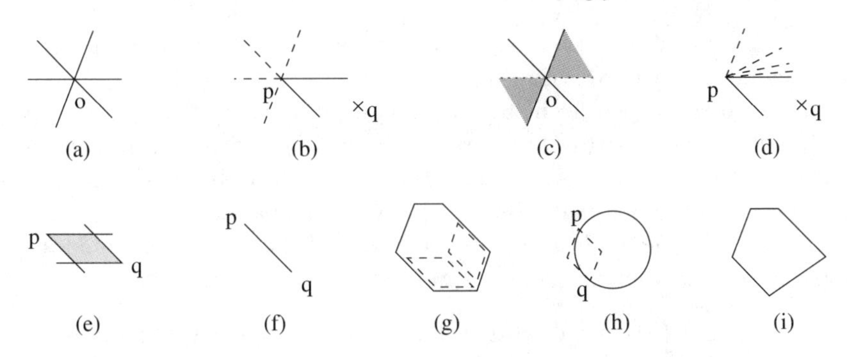

Fig. 2.9. Planar strong $\mathcal{O}$-convexity

2.3 Strongly $\mathcal{O}$-Convex Sets

We now consider an alternative generalization of convexity, called **strong $\mathcal{O}$-convexity,** which also stems from the notion of an orientation set. Rawlins introduced planar strong $\mathcal{O}$-convexity in his doctoral dissertation [35], as part of his research on restricted-orientation visibility. Rawlins and Wood studied the properties of strongly $\mathcal{O}$-convex sets in two dimensions [39,41], and demonstrated that strong $\mathcal{O}$-convexity generalizes not only standard convexity but also the notion of C-oriented polygons [16,18].

The definition of strong $\mathcal{O}$-convexity is based on a characterization of convex sets in terms of visibility. Recall that a set is standard convex if and only if every two of its points are visible to each other. In other words, for every two points of a standard convex set, the line segment joining them is wholly in the set. We introduce a new type of visibility by replacing line segments with different objects, called **$\mathcal{O}$-blocks,** and define strong convexity in terms of this new visibility.

Definition 2.3 ($\mathcal{O}$-blocks). *If the orientation set $\mathcal{O}$ is nonempty, then the* **$\mathcal{O}$-block** *of two points is the intersection of all halfplanes, whose boundaries are $\mathcal{O}$-lines, that contain both points. If $\mathcal{O}$ is empty, then the $\mathcal{O}$-block of any two points is the entire plane.*

To construct the $\mathcal{O}$-block of two points p and q, we draw all $\mathcal{O}$-rays with endpoint p and choose the two that are closest to q, as illustrated in Fig. 2.9b. The two selected rays, with common endpoint p, form the boundary of an angle with vertex p that contains q.

If $\mathcal{O}$ is an infinite set, it may not be closed; thus, we may be unable to choose the ray closest to q. We give an example of a nonclosed orientation set in Fig. 2.9c; all lines in the shaded area are elements of this set, whereas the dotted horizontal line is not in the set. If $\mathcal{O}$ is not closed, we choose two rays with common endpoint p such that, for each of the two selected rays, (1) there is a sequence of $\mathcal{O}$-rays convergent to this ray and (2) there are no $\mathcal{O}$-rays with endpoint p between this ray and the point q, as shown in Fig. 2.9d. The two selected rays again form the boundary of an angle with vertex p.

Similarly, we draw the $\mathcal{O}$-rays from q closest to p and obtain the angle with vertex q whose boundary is formed by these rays. The $\mathcal{O}$-block of p and q is the intersection of the two angles, shown by the shaded parallelogram in Fig. 2.9e. In particular, if the line through p and q is an $\mathcal{O}$-line, then the $\mathcal{O}$-block of p and q is the line segment joining p and q, as shown in Fig. 2.9f.

Definition 2.4 (Strong $\mathcal{O}$-convexity). *A set is* **strongly $\mathcal{O}$-convex** *if, for every two of its points, their $\mathcal{O}$-block is contained in the set.*

We denote the orientation set in Fig. 2.9a by $\mathcal{O}_a$ and that in Fig. 2.9c by $\mathcal{O}_c$. The polygon in Fig. 2.9g is strongly $\mathcal{O}_a$-convex and strongly $\mathcal{O}_c$-convex; two $\mathcal{O}_a$-blocks contained in this polygon are shown by dashed lines. On the other hand, the circle in Fig. 2.9h is neither strongly $\mathcal{O}_a$-convex nor strongly $\mathcal{O}_c$-convex, because the dashed block is not in the circle. Finally, the polygon in Fig. 2.9i is strongly $\mathcal{O}_c$-convex, but not strongly $\mathcal{O}_a$-convex.

Lemma 2.9.

1. *Every translate of a strongly $\mathcal{O}$-convex set is strongly $\mathcal{O}$-convex.*
2. *If C is a collection of strongly $\mathcal{O}$-convex sets, the intersection $\bigcap C$ of these sets is also strongly $\mathcal{O}$-convex.*
3. *For every orientation set $\mathcal{O}$, each strongly $\mathcal{O}$-convex set is standard convex.*
4. *If $\mathcal{O}_1 \subseteq \mathcal{O}_2$, then every strongly $\mathcal{O}_1$-convex set is strongly $\mathcal{O}_2$-convex.*
5. *For two orientation sets $\mathcal{O}_1$ and $\mathcal{O}_2$ through the same point o, strong $\mathcal{O}_1$-convexity is equivalent to strong $\mathcal{O}_2$-convexity if and only if the closure of $\mathcal{O}_1$ is identical to the closure of $\mathcal{O}_2$.*
6. *For a closed orientation set $\mathcal{O}$, a polygon is strongly $\mathcal{O}$-convex if and only if it is convex and its edges are parallel to elements of $\mathcal{O}$.*

Proof.

(1) Since translation preserves $\mathcal{O}$-lines, it also preserves $\mathcal{O}$-blocks, which implies that translates of strongly $\mathcal{O}$-convex sets are strongly $\mathcal{O}$-convex.

(2) If C is a collection of strongly $\mathcal{O}$-convex sets, then, for every two points of the intersection $\bigcap C$, their $\mathcal{O}$-block is a subset of every element of C; hence, this $\mathcal{O}$-block is contained in $\bigcap C$.

(3) For every two points, the line segment joining them is contained in their $\mathcal{O}$-block. Therefore, for every two points of a strongly $\mathcal{O}$-convex set, the segment joining them is wholly in the set.

(4) Suppose that $\mathcal{O}_1 \subseteq \mathcal{O}_2$. The definition of $\mathcal{O}$-blocks readily implies that, for every two points, their $\mathcal{O}_2$-block is a subset of their $\mathcal{O}_1$-block. If P is strongly $\mathcal{O}_1$-convex, then, for every two points of P, their $\mathcal{O}_2$-block is in P, which means that P is strongly $\mathcal{O}_2$-convex.

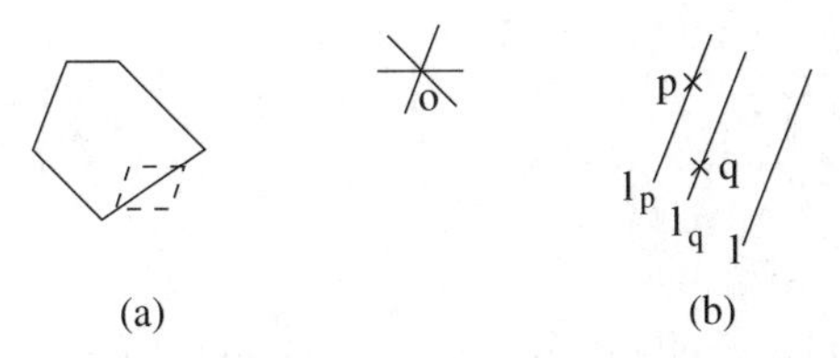

Fig. 2.10. Proof of Lemma 2.9

(5) Let $\mathcal{O}_{\text{cl1}}$ be the closure of $\mathcal{O}_1$, and $\mathcal{O}_{\text{cl2}}$ be the closure of $\mathcal{O}_2$. By definition, the notions of $\mathcal{O}_1$-blocks and $\mathcal{O}_{\text{cl1}}$-blocks are equivalent, which implies that strong $\mathcal{O}_1$-convexity is equivalent to strong $\mathcal{O}_{\text{cl1}}$-convexity. Similarly, strong $\mathcal{O}_2$-convexity is identical to strong $\mathcal{O}_{\text{cl2}}$-convexity. If $\mathcal{O}_{\text{cl1}} = \mathcal{O}_{\text{cl2}}$, then strong $\mathcal{O}_1$-convexity is equivalent to strong $\mathcal{O}_2$-convexity. Suppose, conversely, that $\mathcal{O}_{\text{cl1}} \neq \mathcal{O}_{\text{cl2}}$; without loss of generality, we assume that $\mathcal{O}_{\text{cl1}}$ is not a subset of $\mathcal{O}_{\text{cl2}}$. We consider two distinct points such that the line through them is an $\mathcal{O}_{\text{cl1}}$-line and not an $\mathcal{O}_{\text{cl2}}$-line. Then, the segment joining these two points is strongly $\mathcal{O}_1$-convex but not strongly $\mathcal{O}_2$-convex.

(6) If a polygon P is not convex, it is not strongly $\mathcal{O}$-convex by Part 3 of the proof. If some edge of P is not parallel to any element of $\mathcal{O}$, then, for any two distinct points of this edge, their $\mathcal{O}$-block of is not in P, as shown in Fig. 2.10a; hence, we again conclude that P is not strongly $\mathcal{O}$-convex.

Now suppose that P is a convex polygon and all its edges are parallel to elements of $\mathcal{O}$. Then, P is the intersection of several halfplanes whose boundaries are $\mathcal{O}$-lines. To prove that P is strongly $\mathcal{O}$-convex, we demonstrate that each of these halfplanes is strongly $\mathcal{O}$-convex. Specifically, we show that, for every halfplane whose boundary l is an $\mathcal{O}$-line, and every two points p and q of this halfplane, the $\mathcal{O}$-block of p and q is in the halfplane.

Let l_p be the line through p parallel to l, and l_q be the line through q parallel to l, as shown in Fig. 2.10b. Since l_p and l_q are $\mathcal{O}$-lines, the $\mathcal{O}$-block of p and q is contained in the "strip" between l_p and l_q; hence, this $\mathcal{O}$-block is in the halfplane.

We conclude that P is the intersection of several strongly $\mathcal{O}$-convex halfplanes; therefore, P is strongly $\mathcal{O}$-convex by Part 2 of the proof. □

Finally, we introduce the notion of the **strong $\mathcal{O}$-hull** of a geometric object, which is the intersection of all strongly $\mathcal{O}$-convex sets containing the object; in Fig. 2.11, we show a standard convex hull and a strong $\mathcal{O}$-hull.

Summary

We have introduced two notions of generalized convexity, called $\mathcal{O}$-convexity and strong $\mathcal{O}$-convexity, and presented their basic properties. The important properties of $\mathcal{O}$-convex sets include the separation and decomposition

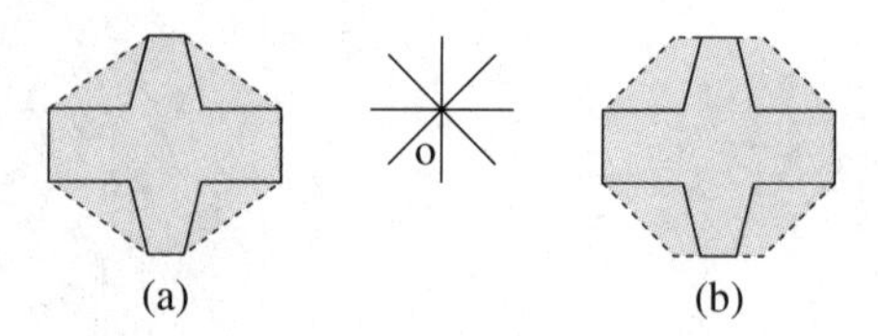

Fig. 2.11. Standard convex hull **(a)** and strong $\mathcal{O}$-hull **(b)**

results (Theorems 2.3 and 2.4), and the characterization of $\mathcal{O}$-convex sets in terms of $\mathcal{O}$-halfplane intersections (Lemma 2.6). The main result for strong $\mathcal{O}$-convexity is the comparison of convexities induced by different orientation sets (Lemma 2.9).

3

Computational Problems

We investigate basic computational tasks in planar strong $\mathcal{O}$-convexity, which include verifying the convexity of a given polygon, computing hulls and kernels, and identifying the regions visible from a given point. Researchers addressed the analogous standard-convexity problems in the early days of computational geometry; for example, consult the text of Preparata and Shamos [34]. They also developed similar techniques for several types of non-traditional convexity, including planar $\mathcal{O}$-convexity.

Rawlins defined planar $\mathcal{O}$-visibility in terms of $\mathcal{O}$-convex curvilinear segments [35]; specifically, two points of a set are $\mathcal{O}$-visible to each other if there is an $\mathcal{O}$-convex curvilinear segment joining them that is wholly in the set. For instance, the points p and q in Fig. 3.1b are $\mathcal{O}$-visible, whereas the points p and x are not. Rawlins, Schuierer and Wood explored the related computational problems and established the following results, which are based on the assumption that $\mathcal{O}$ is a *finite sorted list* of lines.

Proposition 3.1 ($\mathcal{O}$-convexity testing [35, 37]). *If the orientation set $\mathcal{O}$ is a sorted list of m lines, then the time needed to test the $\mathcal{O}$-convexity of a simple n-vertex polygon is $O(n + m)$.*

Proposition 3.2 ($\mathcal{O}$-hull [35,37]). *Suppose that $\mathcal{O}$ is a sorted list of m lines.*

1. *The time needed to compute the $\mathcal{O}$-hull of a simple n-vertex polygon is $O(n + m)$.*
2. *The time needed to compute the $\mathcal{O}$-hull of a set of n points is $O(n \cdot m \cdot \lg n)$.*

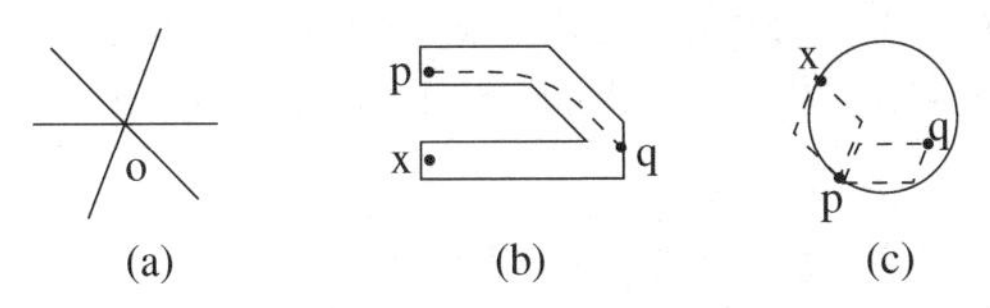

Fig. 3.1. Two types of generalized visibility in the plane

Proposition 3.3 ($\mathcal{O}$-kernel [43, 47]). *Suppose that $\mathcal{O}$ is a sorted list of m lines.*

1. *The time needed to compute the $\mathcal{O}$-kernel of a simple n-vertex polygon is $O(m + n \cdot \lg m)$.*
2. *The time needed to compute the $\mathcal{O}$-kernel of an n-vertex polygon with k holes is $O(n \cdot (\lg n + \lg m) + k \cdot (k + m))$.*

Although Rawlins pointed out similar problems in strong $\mathcal{O}$-convexity [35], he did not pursue this research direction. We now describe methods for solving these problems; most algorithms are based on a reduction to similar problems in standard convexity and visibility.

We define strong $\mathcal{O}$-visibility through $\mathcal{O}$-blocks; specifically, two points of a set are **strongly $\mathcal{O}$-visible** to each other if their $\mathcal{O}$-block is contained in the set. For example, the points p and q in Fig. 3.1c are strongly $\mathcal{O}$-visible, whereas p and x are not; we show the corresponding $\mathcal{O}$-blocks by dashed lines. Note that, if two points are strongly $\mathcal{O}$-visible, they are also standardly visible.

By definition, a set is strongly $\mathcal{O}$-convex if and only if every two of its points are strongly $\mathcal{O}$-visible to each other. Other basic properties of strong $\mathcal{O}$-visibility are also similar to those of standard visibility. In particular, the translation of a set preserves strong $\mathcal{O}$-visibility between its points. Furthermore, if $\mathcal{O}$ includes at least two lines, then every point of a set is strongly $\mathcal{O}$-visible to itself.

We next observe that, if $\mathcal{O}_{\mathrm{cl}}$ is the closure of $\mathcal{O}$, then the $\mathcal{O}$-block of any two points is identical to their $\mathcal{O}_{\mathrm{cl}}$-block, which implies that strong $\mathcal{O}$-visibility is equivalent to strong $\mathcal{O}_{\mathrm{cl}}$-visibility. Thus, *we may restrict attention to the study of strong $\mathcal{O}$-visibility for closed orientation sets.*

We assume that the orientation set $\mathcal{O}$ consists of a finite number of disjoint closed angular intervals; for example, the set in Fig. 3.2a consists of two angular intervals, one of which includes only one line. We also assume that these angular intervals are sorted, which allows binary search for an angular interval that contains a given line. We do not use these assumptions in the study of mathematical properties of strong $\mathcal{O}$-convexity in Sect. 2.3 and Chap. 6, but they are essential for the investigation of computational properties.

We address five basic computational tasks. The first two tasks are to determine whether two points in a polygon are strongly $\mathcal{O}$-visible to each other and whether a polygon is strongly $\mathcal{O}$-convex (Sect. 3.1). The next two problems are to compute the strong $\mathcal{O}$-hull of a point set (Sect. 3.2) and the strong $\mathcal{O}$-kernel of a polygon (Sect. 3.3). Finally, we describe a technique for identifying all points that are strongly $\mathcal{O}$-visible from a given point (Sect. 3.4).

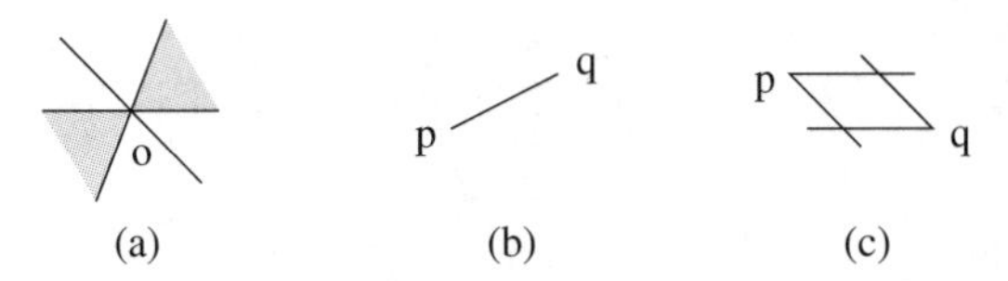

Fig. 3.2. Constructing the $\mathcal{O}$-block of two given points

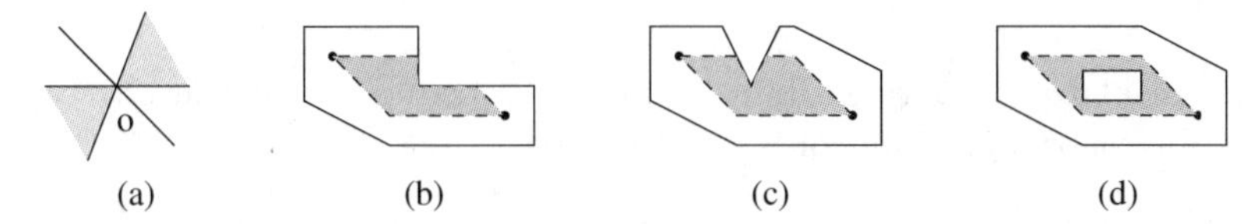

Fig. 3.3. Three cases when points are not strongly $\mathcal{O}$-visible to each other

3.1 Visibility and Convexity Testing

The first problem is to determine whether two points of a polygon are strongly $\mathcal{O}$-visible to each other. The polygon may not be simply connected; that is, it may have holes. To test the strong $\mathcal{O}$-visibility of two points, we construct their $\mathcal{O}$-block and verify its containment in the polygon.

If the line through the two points belongs to an angular interval of $\mathcal{O}$-orientations, then their $\mathcal{O}$-block is a line segment, as shown in Fig. 3.2b. Otherwise, we find the two closest $\mathcal{O}$-orientations and use them to construct the $\mathcal{O}$-block, as shown in Fig. 3.2c. If the orientation set $\mathcal{O}$ comprises m angular intervals, the construction of the $\mathcal{O}$-block of two given points takes $O(\lg m)$ time, because we need to find the two angular-interval boundaries closest to the line through these points. We use the following result to test whether the $\mathcal{O}$-block is inside a given polygon.

Lemma 3.4. *The $\mathcal{O}$-block of two points is wholly in a polygon if and only if the following three conditions hold:*

1. *The vertices of the $\mathcal{O}$-block are in the polygon.*
2. *Every two adjacent vertices of the $\mathcal{O}$-block are standardly visible.*
3. *The polygon does not have holes inside the $\mathcal{O}$-block.*

Proof. We illustrate the violation of each condition in Fig. 3.3. Clearly, if some condition does not hold, the $\mathcal{O}$-block is not contained in the polygon. On the other hand, if the conditions hold, then the polygon's boundary does not intersect the $\mathcal{O}$-block's interior, which implies that the $\mathcal{O}$-block is wholly in the polygon. □

If the polygon has n vertices and the $\mathcal{O}$-block of two given points is a parallelogram, the test for each condition of Lemma 3.4 takes $O(n)$ time. If the $\mathcal{O}$-block of the given points is a line segment, then their strong $\mathcal{O}$-visibility

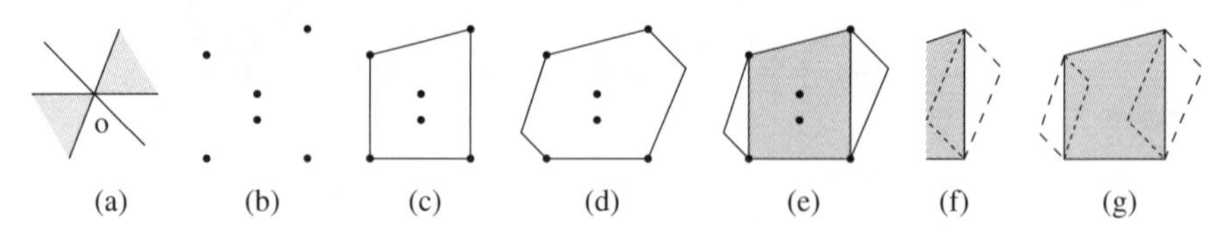

Fig. 3.4. Construction of the strong $\mathcal{O}$-hull

is equivalent to standard visibility, which can also be verified in $O(n)$ time. Thus, the following result holds in both cases.

Theorem 3.5. *If the orientation set $\mathcal{O}$ is a sorted list of m disjoint angular intervals, then the time needed to test the strong $\mathcal{O}$-visibility of two points in an n-vertex polygon is $O(n + \lg m)$.*

Next, we describe an algorithm for verifying the strong $\mathcal{O}$-convexity of a given polygon, based on the condition stated in Lemma 2.9; specifically, a polygon is strongly $\mathcal{O}$-convex if and only if it is convex and all its edges are parallel to elements of $\mathcal{O}$. Testing whether a polygon with n vertices is standard convex takes $O(n)$ time. If it is convex, then the orientations of its edges are in sorted order, and we need to check whether each orientation in the sorted list of edge orientations belongs to one of the angular intervals in $\mathcal{O}$, where the angular intervals are also in sorted order.

If $m \leq n$, we can test whether a sorted list of n orientations is a subset of a union of m angular intervals in $O(n)$ time; if $m > n$, an efficient test takes $O(n \cdot \lg \frac{n+m}{n})$ time. We express the running time as $O(n + n \cdot \lg \frac{n+m}{n})$, which is equivalent to $O(n)$ for $m \leq n$ and to $O(n \cdot \lg \frac{n+m}{n})$ for $m > n$.

Theorem 3.6. *If $\mathcal{O}$ is a sorted list of m disjoint angular intervals, the time of testing the strong $\mathcal{O}$-convexity of an n-vertex polygon is $O(n + n \cdot \lg \frac{n+m}{n})$.*

3.2 Strong $\mathcal{O}$-Hull

Recall that the strong $\mathcal{O}$-hull of a geometric object is the intersection of all strongly $\mathcal{O}$-convex sets containing the object. For example, consider the orientation set in Fig. 3.4a and the set of six points in Fig. 3.4b. We show the standard convex hull of this six-point set in Fig. 3.4c and its strong $\mathcal{O}$-hull in Fig. 3.4d. The computation of the strong $\mathcal{O}$-hull is based on the following observation.

Lemma 3.7. *The strong $\mathcal{O}$-hull of a geometric object is identical to the strong $\mathcal{O}$-hull of the standard convex hull of the object.*

Proof. Let P be a geometric object, Q be its standard convex hull, and S-hull(P) and S-hull(Q) be their strong $\mathcal{O}$-hulls. For example, if P is the six-point set in Fig. 3.4e, then Q is the shaded polygon, and S-hull(P) is the outer hexagon. By the definition of convex hulls, P is a subset of Q, which implies that S-hull(P) $\subseteq$ S-hull(Q). On the other hand, since S-hull(P) is standard convex, and Q is the *minimal* standard convex set containing P, we conclude that $Q \subseteq$ S-hull(P), and therefore S-hull(Q) $\subseteq$ S-hull(P). These two opposite inclusions imply that S-hull(Q) = S-hull(P). □

The first step in constructing the strong $\mathcal{O}$-hull of an n-point set is finding its standard convex hull, which takes $O(n \cdot \lg n)$ time [12]. The resulting hull is a convex polygon with at most n vertices, and the next step is finding the strong $\mathcal{O}$-hull of this polygon.

For every two adjacent vertices of the polygon, we determine their $\mathcal{O}$-block and identify the half of the $\mathcal{O}$-block outside of the polygon, called the **outer halfblock**. We illustrate this construction in Fig. 3.4f, where the shaded region is the polygon's interior, the dashed lines show the outer halfblock of two adjacent vertices, and the dotted lines mark the other half of their $\mathcal{O}$-block. The concatenated boundaries of the outer halfblocks form the contour that bounds the strong $\mathcal{O}$-hull of the polygon, as shown in Fig. 3.4g.

The construction of every $\mathcal{O}$-block requires finding the two $\mathcal{O}$-orientations closest to the corresponding edge, which takes $O(\lg m)$ time. The overall time needed to compute the standard convex hull of the point set and to construct the $\mathcal{O}$-blocks is $O(n \cdot (\lg n + \lg m))$.

Observe that the orientations of the edges of the standard convex hull are in sorted order, which allows us to reduce the time of finding the two closest $\mathcal{O}$-orientations for each of the edges. Specifically, we can construct the $\mathcal{O}$-blocks in $O(n + n \cdot \lg \frac{n+m}{n})$ time. Unfortunately, it does not improve the overall running time, because computing the standard convex hull takes $O(n \cdot \lg n)$ time, which yields a total running time of $O(n \cdot (\lg n + \lg m))$.

The next observation implies that the same algorithm can compute the strong $\mathcal{O}$-hull of a polygon; therefore, the time needed to compute a polygon's strong $\mathcal{O}$-hull is also $O(n \cdot (\lg n + \lg m))$.

Lemma 3.8. *The strong $\mathcal{O}$-hull of a polygon is identical to the strong $\mathcal{O}$-hull of the point set formed by the polygon's vertices.*

Proof. Clearly, the standard convex hull of a polygon is identical to the standard convex hull of its vertices; for example, consult the text of Grünbaum, Klee and Perles [15]. Furthermore, the strong $\mathcal{O}$-hull of a point set is identical to the strong $\mathcal{O}$-hull of the standard convex hull of the set by Lemma 3.7, which immediately implies the desired result. □

If a polygon is simple, then we can compute its standard convex hull in $O(n)$ time [27], thus reducing the overall time needed to construct the

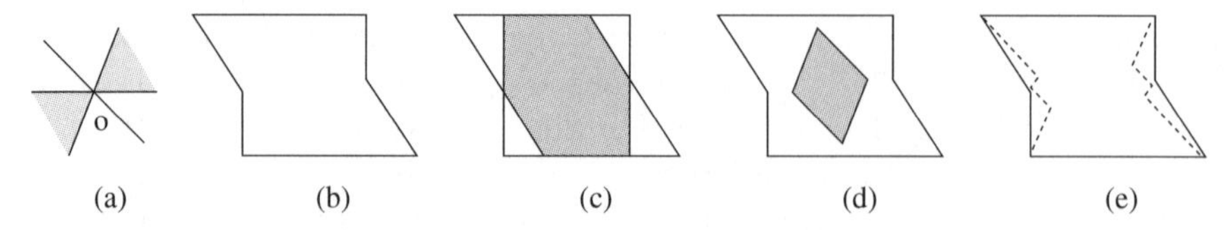

Fig. 3.5. Standard kernel, strong $\mathcal{O}$-kernel and inner halfblocks of a polygon

strong $\mathcal{O}$-hull to $O(n + n \cdot \lg \frac{n+m}{n})$. We state the results for the computation of the strong $\mathcal{O}$-hull as a theorem.

Theorem 3.9. *Suppose that $\mathcal{O}$ is a sorted list of m disjoint angular intervals.*

1. *The time needed to compute the strong $\mathcal{O}$-hull of an n-point set or that of an n-vertex polygon is $O(n \cdot (\lg n + \lg m))$.*
2. *The time needed to compute the strong $\mathcal{O}$-hull of a simple polygon with n vertices is $O(n + n \cdot \lg \frac{n+m}{n})$.*

3.3 Strong $\mathcal{O}$-Kernel

The standard **kernel** of a geometric object is the set of points that are standardly visible from all points of the object. For example, the polygon in Fig. 3.5b has a nonempty standard kernel, shown as a shaded region in Fig. 3.5c. Note that only simply connected objects can have nonempty standard kernels, and that an object is standard convex if and only if its standard kernel is identical to the object itself. If an object has a nonempty standard kernel, it is called a **star-shaped object**. The computation of the standard kernel of a polygon with n vertices takes $O(n)$ time [23].

The **strong $\mathcal{O}$-kernel** is the set of all points that are strongly $\mathcal{O}$-visible from all points of the object. In Fig. 3.5d, we show the strong $\mathcal{O}$-kernel of the same polygon, for the orientation set in Fig. 3.5a. The problem of finding the strong $\mathcal{O}$-kernel of a polygon is reducible to the standard-kernel computation. We derive the properties of the strong $\mathcal{O}$-kernel used in the reduction and then describe the algorithm.

First, observe that a point of a polygon belongs to the standard kernel if and only if it is standardly visible from every vertex; for example, consult the text of Preparata and Shamos [34]. The analogous result holds for strong $\mathcal{O}$-visibility.

Lemma 3.10. *A point of a polygon is in the strong $\mathcal{O}$-kernel if and only if it is strongly $\mathcal{O}$-visible from all vertices of the polygon.*

Proof. Clearly, if a point is in the strong $\mathcal{O}$-kernel, then it is strongly $\mathcal{O}$-visible from all vertices. To prove the converse, we consider a point p that is strongly

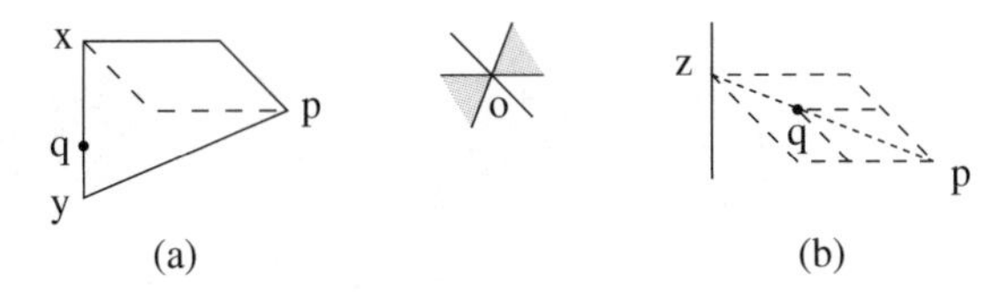

Fig. 3.6. Proof of Lemma 3.10

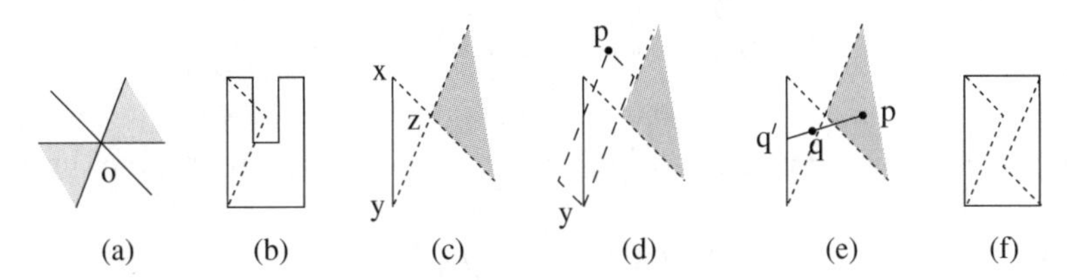

Fig. 3.7. Proof of Lemma 3.11

$\mathcal{O}$-visible from all vertices, and show that p is strongly $\mathcal{O}$-visible from every point q of the polygon.

First, suppose that q is on some edge, whose endpoints are denoted by x and y, as shown in Fig. 3.6a. Note that the $\mathcal{O}$-block of p and x is in the polygon, and the $\mathcal{O}$-block of p and y is also in the polygon; therefore, the contour formed by these $\mathcal{O}$-blocks and the edge, which is marked by solid lines in Fig. 3.6a, is completely in the polygon. Clearly, the $\mathcal{O}$-block of p and q is inside this contour, which implies that p is strongly $\mathcal{O}$-visible from q.

Now suppose that q is in the polygon's interior. We draw the line from p to q, extending it until it intersects the boundary, and consider the point z of the intersection, as shown in Fig. 3.6b. Since p is strongly $\mathcal{O}$-visible from z and the $\mathcal{O}$-block of p and q is inside the $\mathcal{O}$-block of p and z, we conclude that p is strongly $\mathcal{O}$-visible from q. □

The first step in the construction of the strong $\mathcal{O}$-kernel is to find the inner halfblocks of a polygon, which are analogous to the outer halfblocks. Specifically, the **inner halfblock** of two adjacent vertices is the half of their $\mathcal{O}$-block opposite to the outer halfblock. We show the inner halfblocks of the polygon in Fig. 3.4g by dotted lines. Note that an inner halfblock may not be completely in the polygon, as illustrated in Fig. 3.7b.

Lemma 3.11. *If a polygon has a pair of adjacent vertices whose inner halfblock is not in the polygon, then the polygon's strong $\mathcal{O}$-kernel is empty.*

Proof. Let x and y be adjacent vertices, whose inner halfblock is not completely in the polygon, and let z be the third vertex of their halfblock, as shown in Fig. 3.7c. Suppose that the strong $\mathcal{O}$-kernel of the polygon is nonempty and p is one of its points. Note that p cannot be above the line through y and z, because then the $\mathcal{O}$-block of p and y would not be completely in the

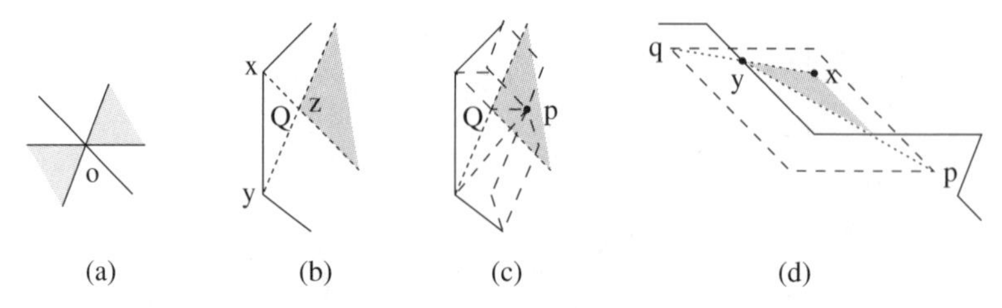

Fig. 3.8. Proofs of Lemmas 3.12 and 3.13

polygon, as shown in Fig. 3.7d. Similarly, p cannot be below the line through x and z, which implies that p is in the shaded angle.

We now pick a point q that is in the halfblock and not in the polygon, and denote the intersection of the line through p and q with the edge by q', as shown in Fig. 3.7e. Then, p is not strongly $\mathcal{O}$-visible from q', which yields a contradiction. □

The converse of Lemma 3.11 does not hold; that is, a polygon's strong $\mathcal{O}$-kernel may be empty even if every inner halfblock is in the polygon. For example, the strong $\mathcal{O}$-kernel of the polygon in Fig. 3.7f is empty.

Lemma 3.12. *Suppose that P_1 is a polygon and $Q \subseteq P_1$ is the inner halfblock of some pair of adjacent vertices, and that we construct a polygon P_2 by cutting Q from P_1; that is, P_2 is the closure of $P_1 - Q$. Then, the strong $\mathcal{O}$-kernel of P_2 is identical to the strong $\mathcal{O}$-kernel of P_1.*

Proof. We first show that every point p of P_2's strong $\mathcal{O}$-kernel is in the strong $\mathcal{O}$-kernel of P_1. If p is in the strong $\mathcal{O}$-kernel of P_2, it is strongly $\mathcal{O}$-visible from all vertices of P_2. Since every vertex of P_1 is a vertex of P_2, and P_1 is a superset of P_2, we conclude that p is strongly $\mathcal{O}$-visible from all vertices of P_1; therefore, p is in the strong $\mathcal{O}$-kernel of P_1 by Lemma 3.10.

We next show that, conversely, every point p of P_1's strong $\mathcal{O}$-kernel is in the strong $\mathcal{O}$-kernel of P_2. Consider adjacent vertices x and y, which give rise to the inner halfblock Q, and let z be the third vertex of Q, as illustrated in Fig. 3.8b. We have shown, in the proof of Lemma 3.11, that p is in the shaded angle. Therefore, for every vertex q of P_2, the $\mathcal{O}$-block of p and q is either above the line through x and z or below the line through y and z, as shown in Fig. 3.8c, which means that the $\mathcal{O}$-block of p and q does not intersect the interior of Q. We conclude that p is strongly $\mathcal{O}$-visible from all vertices of P_2; hence, it is in the strong $\mathcal{O}$-kernel of P_2 by Lemma 3.10. □

Lemma 3.13. *If all edges of a polygon are parallel to elements of $\mathcal{O}$, then its strong $\mathcal{O}$-kernel is identical to its standard kernel.*

Proof. Clearly, if a point p is in the strong $\mathcal{O}$-kernel of the polygon, then it is in the standard kernel. To show the converse, suppose that p is not in the strong $\mathcal{O}$-kernel, and q is a point that is not strongly $\mathcal{O}$-visible from p, as shown in Fig. 3.8d. We pick a point x that is in the $\mathcal{O}$-block of p and q, and not in the polygon. Let y be the first intersection of the line segment from x to q with the polygon's boundary, shown by solid lines in Fig. 3.8d. Since the polygon's edge through y is parallel to some element of $\mathcal{O}$, it does not intersect the shaded angle, except at its vertex y. Thus, y is not standardly visible from p, which implies that p is not in the standard kernel. □

To find the strong $\mathcal{O}$-kernel of a polygon, we construct the inner halfblock for every edge, as shown in Fig. 3.5e. The catenation of the halfblock boundaries forms a new polygon, whose standard kernel is identical to the strong $\mathcal{O}$-kernel of the original polygon. If some halfblocks are not contained in the original polygon, then the new polygon is not simple, and its strong $\mathcal{O}$-kernel is empty. If all halfblocks are in the original polygon, then the new polygon is simple, and the computation of its standard kernel takes $O(n)$ time. The construction of each inner halfblock takes $O(\lg m)$ time, which leads to the following result.

Theorem 3.14. *If $\mathcal{O}$ is a sorted list of m disjoint angular intervals, the time needed to compute the strong $\mathcal{O}$-kernel of an n-vertex polygon is $O(n \cdot \lg m)$.*

3.4 Visibility from a Point

Suppose that a finite set of "obstacles" obstructs visibility in the plane, and two points are considered strongly $\mathcal{O}$-visible to each other if their $\mathcal{O}$-block does not intersect any obstacles. We give an algorithm for finding all points that are strongly $\mathcal{O}$-visible from a given spot.

As a first step, suppose that all obstacles are *points,* and consider the problem of finding *all obstacle points* that are strongly $\mathcal{O}$-visible from a given point p. We illustrate this problem in Fig. 3.9b, where small squares mark the obstacle points visible from p, and small triangles show the invisible obstacles.

Consider the angles formed by the $\mathcal{O}$-lines through p, shown in Fig. 3.9c, and observe that an obstacle point affects the visibility from p only within the angle that contains this point. In particular, if an obstacle is inside an angular interval of $\mathcal{O}$-lines, shown by the shaded angles in Fig. 3.9c, it obstructs the visibility only along the line through p and the obstacle.

If a point is in an angle between two angular intervals, then it obstructs part of the angle, and does not affect the visibility from p in other angles; thus, we can solve the visibility problem separately for each angle. We use the sides of an angle as coordinate axes, and assign x and y coordinates to each obstacle in the angle, as shown in Fig. 3.9d. Then, an obstacle q_2 is not visible from p

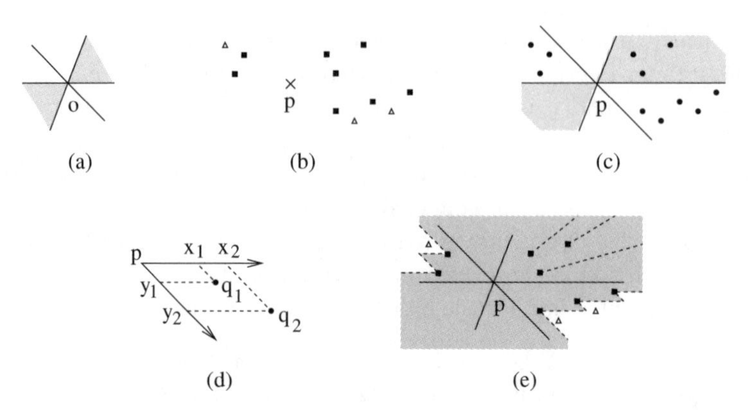

Fig. 3.9. Construction of a strong $\mathcal{O}$-visibility polygon for point obstacles

Let $q_1, q_2, ..., q_k$ be the obstacles in the angle, sorted on x-coordinates, and $y_1, y_2, ..., y_k$ be their y-coordinates.

$\mathit{Visible} := \{q_1\}$ (set of visible obstacle points)
$y_{\min} := y_1$ (minimal y among the processed obstacles)
for $i := 2$ **to** k **do**
    if $y_i < y_{\min}$
        then $\mathit{Visible} := \mathit{Visible} \cup \{q_i\}$
            $y_{\min} := y_i$
return *Visible*

Fig. 3.10. Finding the obstacles strongly $\mathcal{O}$-visible from p, for one of the angles

if and only if there is an obstacle q_1 such that $x_1 \leq x_2$ and $y_1 \leq y_2$. This observation readily leads to an algorithm for identifying the visible obstacles given in Fig. 3.10.

The algorithm sorts the obstacle points, contained in an angle, in increasing order of their x-coordinates. If several points have the same x-coordinate, it sorts these points in increasing order of their y-coordinates. Then, the algorithm processes the obstacle points in sorted order. If an obstacle's y-coordinate is strictly smaller than the y-coordinates of all previously processed obstacles, then it is visible from p. A convenient visualization of this algorithm is a sweep of a parallel-to-y line through the obstacles, in the direction of increasing x, as illustrated in Fig. 3.11.

The top-level algorithm groups the obstacles by angle and then calls the sweep algorithm for each angle. If the total number of obstacles is n, then finding all angles that contain at least one obstacle takes $O(n \cdot \lg m)$ time, grouping the obstacles by angle and sorting them within each angle takes $O(n \cdot \lg n)$ time, and applying the sweep algorithm to all angles takes $O(n)$ time. Thus, the overall time needed to identify the obstacles visible from p is $O(n \cdot (\lg n + \lg m))$.

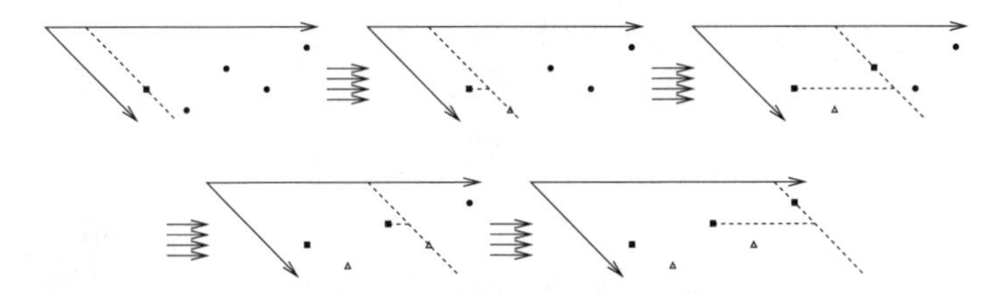

Fig. 3.11. Sweep-line view of the strong $\mathcal{O}$-visibility computation

We can adapt this algorithm to compute the set of *all* points that are strongly $\mathcal{O}$-visible from p. We show the boundary of this set in Fig. 3.9e; the boundary includes (1) "invisible" rays in the angular intervals of $\mathcal{O}$-lines through p and (2) "stair-shaped" polygonal lines through the visible obstacles in the angles between angular intervals, one polygonal line for each angle. The algorithm first identifies the invisible rays and then constructs the stair-shaped lines. It identifies these lines during the sweeps, shown in Fig. 3.11, which does not increase the overall sweep time.

The resulting set of visible points is an unbounded star-shaped polygon, shown by the shaded area in Fig. 3.9e, which is called the **strong $\mathcal{O}$-visibility polygon** of p. This polygon is open; that is, it does not include its boundary. The next result enables us to extend this construction to *line-segment* obstacles.

Lemma 3.15. *Suppose that $Obst_1$ is a set of line-segment obstacles, and $Obst_2$ is the obstacle set formed by the endpoints of these line segments. Then, two points are strongly $\mathcal{O}$-visible to each other with respect to $Obst_1$ if and only if they are strongly $\mathcal{O}$-visible with respect to $Obst_2$ and standardly visible with respect to $Obst_1$.*

Proof. Clearly, if points p and q are strongly $\mathcal{O}$-visible with respect to $Obst_1$, then they are strongly $\mathcal{O}$-visible for $Obst_2$ and standardly visible for $Obst_1$, as shown in Fig. 3.12b. If p and q are not strongly $\mathcal{O}$-visible with respect to $Obst_1$, then either (1) the endpoint of some line-segment obstacle is in their $\mathcal{O}$-block, which means that they are not strongly $\mathcal{O}$-visible with respect to $Obst_2$, as shown in Fig. 3.12c, or (2) one of the line-segment obstacles cuts through the $\mathcal{O}$-block and obstructs the standard visibility between p and q, as shown in Fig. 3.12d. □

The algorithm for constructing the strong $\mathcal{O}$-visibility polygon of a point p, with respect to a set $Obst_1$ of n line-segment obstacles, consists of three steps; we illustrate these steps for the obstacle set in Fig. 3.13b. First, the algorithm identifies the set $Obst_2$ of the obstacle endpoints, and finds the strong $\mathcal{O}$-visibility polygon with respect to $Obst_2$, as shown in Fig. 3.13c. Second, it constructs the standard visibility polygon with respect to $Obst_1$, as shown

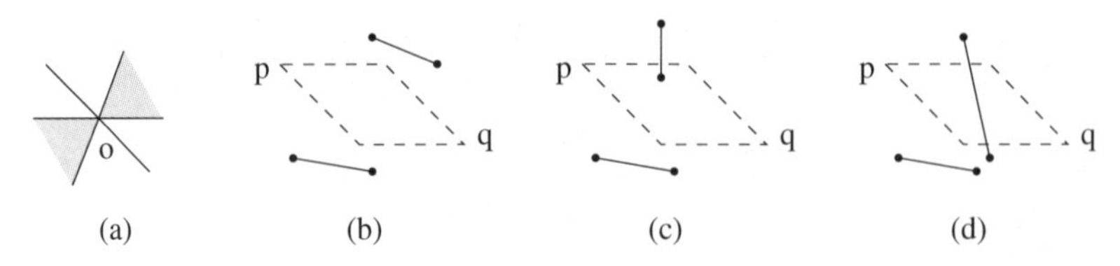

Fig. 3.12. Proof of Lemma 3.15

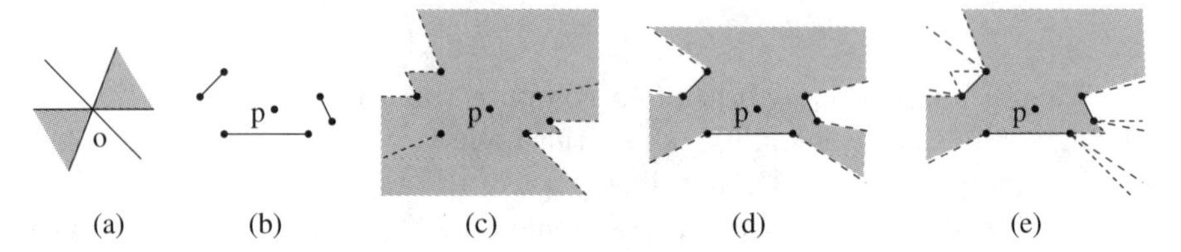

Fig. 3.13. Construction of a strong $\mathcal{O}$-visibility polygon for line-segment obstacles

in Fig. 3.13d, which takes $O(n \cdot \lg n)$ time [22]. Finally, the third step is to compute the intersection of the two polygons, as shown in Fig. 3.13e. Most linear-time algorithms for finding the intersection of convex polygons [34] are readily applicable to the intersection of two star-shaped polygons with a common kernel point. Since the two visibility polygons have the common kernel point p, we can compute their intersection in $O(n)$ time.

Theorem 3.16. *If $\mathcal{O}$ is a sorted list of m disjoint angular intervals, and the obstacle set comprises n line segments, then the time needed to compute the strong $\mathcal{O}$-visibility polygon of a point is $O(n \cdot (\lg n + \lg m))$.*

Finally, note that we can use the same algorithm for polygonal obstacles and for visibility inside a polygon, since these problems are readily reducible to visibility with respect to line-segment obstacles.

Summary

We have begun exploration of the computational properties of strong $\mathcal{O}$-convexity and strong $\mathcal{O}$-visibility, and developed algorithms for computing the strong $\mathcal{O}$-hull of a point set, the strong $\mathcal{O}$-kernel of a polygon, and the strong $\mathcal{O}$-visibility polygon of a point surrounded by line-segment obstacles. The dependency between the size m of the orientation set and the running time is logarithmic, which means that the computation is efficient for large m.

In Table 3.1, we compare the running times of the developed algorithms with analogous results for standard visibility and $\mathcal{O}$-visibility; however, note that $\mathcal{O}$-visibility algorithms are somewhat less general. Specifically, they are for orientation sets with a finite number of lines, whereas the results on strong $\mathcal{O}$-convexity are for sets with a finite number of angular intervals.

Table 3.1. Worst-case time complexity for the three notions of visibility

	Standard visibility	$\mathcal{O}$-visibility	Strong $\mathcal{O}$-visibility
Visibility of two points	n	unknown	$n + \lg m$
Convexity of a polygon	n	$n + m$	$n + \lg \frac{n+m}{n}$
Hull of a point set	$n \cdot \lg n$	$n \cdot m \cdot \lg n$	$n \cdot (\lg n + \lg m)$
Hull of a simple polygon	n	$n + m$	$n + n \cdot \lg \frac{n+m}{n}$
Kernel of a simple polygon	n	$m + n \cdot \lg m$	$n \cdot \lg m$
Visibility from a point	$n \cdot \lg n$	unknown	$n \cdot (\lg n + \lg m)$

Since we have not investigated the lower bounds for the running times, there is the possibility of improving the efficiency of some algorithms. Other closely related problems include dynamic maintenance of the strong $\mathcal{O}$-hull, construction of maximal strongly $\mathcal{O}$-convex subsets of a given polygon and dynamic maintenance of the strong $\mathcal{O}$-visibility polygon.

4

Higher Dimensions

We now extend the notion of $\mathcal{O}$-convexity to a d-dimensional space $\mathcal{R}^d$. First, we define orientation sets in higher dimensions (Sect. 4.1), consider $\mathcal{O}$-convex sets in $\mathcal{R}^d$, and introduce $\mathcal{O}$-connected sets, which are a subclass of $\mathcal{O}$-convex sets with several special properties (Sect. 4.2). Then, we explore properties of $\mathcal{O}$-connected curves (Sect. 4.3) and present visibility results for $\mathcal{O}$-convex and $\mathcal{O}$-connected sets (Sect. 4.4).

4.1 Orientation Sets

We introduce a set $\mathcal{O}$ of hyperplanes through a fixed point o, show how it gives rise to $\mathcal{O}$-lines, and then define higher-dimensional $\mathcal{O}$-convex sets in terms of their intersections with $\mathcal{O}$-lines.

A **hyperplane** in d dimensions is a subset of $\mathcal{R}^d$ that is a $(d-1)$-dimensional space; for example, hyperplanes in three dimensions are the usual planes. Analytically, a hyperplane is a set of points satisfying a linear equation $a_1x_1 + a_2x_2 + \cdots + a_dx_d = b$ in Cartesian coordinates.

A hyperplane is a special case of a **flat**, which is a subset of $\mathcal{R}^d$ that is itself a k-dimensional space, where $k \leq d$. For example, points, lines and two-dimensional planes are flats, and the whole space $\mathcal{R}^d$ is also a flat. Analytically, a k-dimensional flat is represented by a system of $d-k$ independent linear equations. We use the following properties of flats.

Proposition 4.1 (Properties of flats).

1. *The intersection of flats is either empty or a flat.*
2. *The intersection of a k-dimensional flat and a hyperplane is empty, the k-dimensional flat itself or a $(k-1)$-dimensional flat.*

Two flats are **parallel** if they are translates of each other; note that parallel flats are of the same dimension. In particular, hyperplanes are parallel if they differ only by the value of b in the equation $a_1x_1 + a_2x_2 + \cdots + a_dx_d = b$.

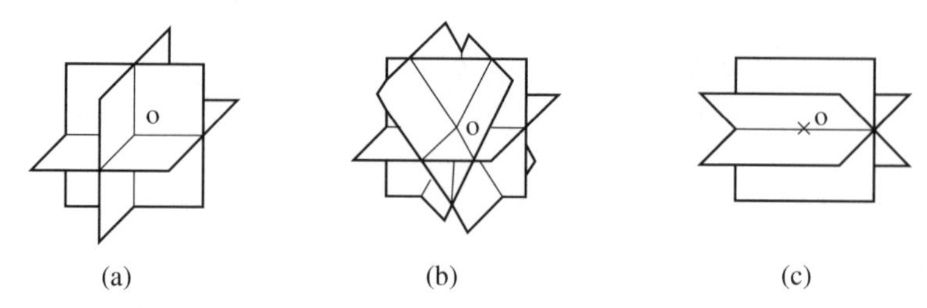

Fig. 4.1. Finite orientation sets

Definition 4.1 (Orientation sets and $\mathcal{O}$-hyperplanes). *An* **orientation set** $\mathcal{O}$ *is a set of hyperplanes through a fixed point o. A hyperplane parallel to one of the elements of* $\mathcal{O}$ *is called an* $\mathcal{O}$**-hyperplane.**

Note that every translate of an $\mathcal{O}$-hyperplane is an $\mathcal{O}$-hyperplane, which implies that a particular choice of the point o is unimportant. When we consider several different orientation sets in $\mathcal{R}^d$, we assume that the elements of these sets are through the same common point o.

In Fig. 4.1, we give examples of finite orientation sets in three dimensions. The first set contains three mutually orthogonal planes; we call it an **orthogonal-orientation set.** The second set consists of four planes, and the third set comprises three planes that have a common horizontal line.

The $\mathcal{O}$**-lines** in $\mathcal{R}^d$ are formed by the intersections of $\mathcal{O}$-hyperplanes; that is, a line is an $\mathcal{O}$-line if it is the intersection of several $\mathcal{O}$-hyperplanes. Note that, since every $\mathcal{O}$-hyperplane is parallel to some element of the orientation set $\mathcal{O}$, every $\mathcal{O}$-line is parallel to some line formed by the intersection of several elements of $\mathcal{O}$. For example, the intersections of the four elements of the orientation set in Fig. 4.1b form six lines through o, and every $\mathcal{O}$-line for this orientation set is parallel to one of these six lines.

When $\mathcal{O}$ is nonempty and the intersection of the elements of $\mathcal{O}$ is the point o, rather than a proper superset of o, we say that $\mathcal{O}$ has the **point-intersection property.** For example, the orientation sets in Fig. 4.1a,b have this property, whereas the set in Fig. 4.1c does not. Some results hold only for orientation sets that satisfy the point-intersection property.

Lemma 4.2. *If an orientation set* $\mathcal{O}$ *has the point-intersection property, then:*

1. *There is at least one* $\mathcal{O}$*-line.*
2. *For every line, there is an* $\mathcal{O}$*-hyperplane that intersects it and does not contain it.*

Proof.

(1) We consider a minimal set of $\mathcal{O}$-hyperplanes whose intersection forms o, and denote these hyperplanes by $\mathcal{H}_1, \mathcal{H}_2, \ldots, \mathcal{H}_n$. Then, $\mathcal{H}_2 \cap \cdots \cap \mathcal{H}_n$

is a flat different from the point o, and the intersection of this flat with $\mathcal{H}_1$ is o, which implies that $\mathcal{H}_2 \cap \cdots \cap \mathcal{H}_n$ is a line; by definition, it is an $\mathcal{O}$-line.

(2) We consider a line l and assume, without loss of generality, that l is through o. The intersection of all elements of $\mathcal{O}$ is o; therefore, for some hyperplane $\mathcal{H}$ of $\mathcal{O}$, the line l is not contained in $\mathcal{H}$. □

If the orientation set $\mathcal{O}$ does not satisfy the point-intersection property, then the intersection of the elements of $\mathcal{O}$ is either a line or a higher-dimensional flat. If this intersection is a line, as shown in Fig. 4.1c, then there is exactly one $\mathcal{O}$-line through o and all other $\mathcal{O}$-lines are parallel to it. If the intersection is neither a point nor a line, then there are no $\mathcal{O}$-lines at all.

We next consider the flats formed by the intersections of $\mathcal{O}$-hyperplanes.

Definition 4.2 ($\mathcal{O}$-flats). *A flat formed by the intersection of several $\mathcal{O}$-hyperplanes is called an* **$\mathcal{O}$-flat**. *The $\mathcal{O}$-hyperplanes themselves and the whole space $\mathcal{R}^d$ are also considered $\mathcal{O}$-flats.*

Since every $\mathcal{O}$-hyperplane is parallel to some element of the orientation set $\mathcal{O}$, every $\mathcal{O}$-flat is parallel to some flat formed by the intersection of several elements of $\mathcal{O}$. In particular, if the point o is the intersection of several elements of $\mathcal{O}$, then every point in $\mathcal{R}^d$ is an $\mathcal{O}$-flat. For example, the orthogonal-orientation set in three dimensions, shown in Fig. 4.1a, gives rise to the following $\mathcal{O}$-flats through o: the whole space, the three mutually orthogonal $\mathcal{O}$-planes, the three $\mathcal{O}$-lines formed by the intersections of these planes and the point o.

The next result readily follows from the definition of $\mathcal{O}$-flats.

Lemma 4.3.

1. *Every translate of an $\mathcal{O}$-flat is an $\mathcal{O}$-flat.*
2. *The intersection of $\mathcal{O}$-flats is either empty or an $\mathcal{O}$-flat.*

We next describe lower-dimensional orientation sets contained in $\mathcal{O}$-flats. We consider an $\mathcal{O}$-flat η of dimension k; we treat η as an independent k-dimensional space and define the orientation set $\mathcal{O}_\eta$ and $\mathcal{O}_\eta$-flats in this space.

An **$\mathcal{O}_\eta$-flat** is an $\mathcal{O}$-flat contained in η; the $(k-1)$-dimensional $\mathcal{O}_\eta$-flats ("$\mathcal{O}_\eta$-hyperplanes") through some fixed point o_η form the lower-dimensional orientation set $\mathcal{O}_\eta$. For instance, consider the orientation set in Fig. 4.2a. The $\mathcal{O}$-plane η contains vertical and horizontal $\mathcal{O}$-lines; hence, the lower-dimensional set $\mathcal{O}_\eta$ comprises the vertical and horizontal lines through o_η. In Fig. 4.2b, we give another example of a lower-dimensional orientation set.

The next result implies that $\mathcal{O}_\eta$-flats have all necessary properties of $\mathcal{O}$-flats. Furthermore, if $\mathcal{O}$ has the point-intersection property, then $\mathcal{O}_\eta$ also has this property.

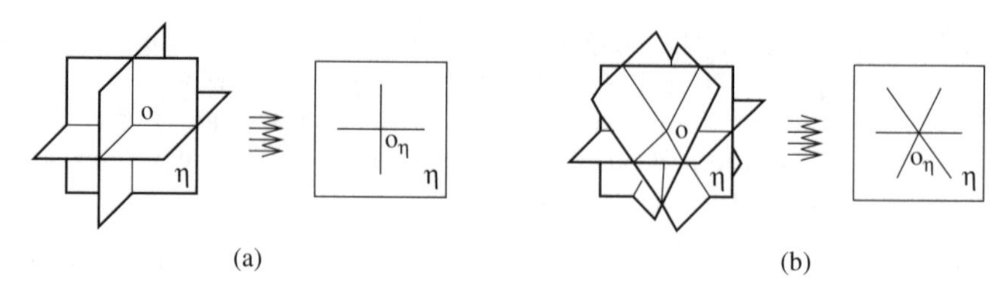

Fig. 4.2. Lower-dimensional orientation sets

Lemma 4.4. *Let η be an $\mathcal{O}$-flat of dimension k.*

1. *Every translate of an $\mathcal{O}_\eta$-flat within the space η is an $\mathcal{O}_\eta$-flat.*
2. *A set $H \subseteq \eta$ is an $\mathcal{O}_\eta$-flat if and only if it is either η itself or the intersection of several $(k-1)$-dimensional $\mathcal{O}_\eta$-flats.*
3. *If the orientation set $\mathcal{O}$ has the point-intersection property, then $\mathcal{O}_\eta$ also has this property; that is, $\mathcal{O}_\eta$ is nonempty and $\bigcap \mathcal{O}_\eta = o_\eta$.*

Proof.

(1) Every $\mathcal{O}_\eta$-flat is an $\mathcal{O}$-flat; hence, a translate of an $\mathcal{O}_\eta$-flat is also an $\mathcal{O}$-flat. If the translate is contained in η, then it is an $\mathcal{O}_\eta$-flat.

(2) Since $\mathcal{O}_\eta$-flats are $\mathcal{O}$-flats, the intersection of $(k-1)$-dimensional $\mathcal{O}_\eta$-flats is an $\mathcal{O}$-flat contained in η, which implies that it is an $\mathcal{O}_\eta$-flat. Next, we show that every $\mathcal{O}_\eta$-flat H distinct from η is the intersection of $(k-1)$-dimensional $\mathcal{O}_\eta$-flats. Since H is an $\mathcal{O}$-flat, it is formed by the intersection of several $\mathcal{O}$-hyperplanes, say $\mathcal{H}_1, \mathcal{H}_2, \ldots, \mathcal{H}_n$. Let $\mathcal{H}_1, \mathcal{H}_2, \ldots, \mathcal{H}_k$ be the hyperplanes among them that do not contain η. Then, $\mathcal{H}_1 \cap \eta, \mathcal{H}_2 \cap \eta, \ldots, \mathcal{H}_k \cap \eta$ are $(k-1)$-dimensional flats by Proposition 4.1; thus, they are $\mathcal{O}_\eta$-flats and their intersection forms H.

(3) We assume, without loss of generality, that $o_\eta = o$. If the intersection of the elements of $\mathcal{O}$ is the point o, then there is some $\mathcal{O}$-hyperplane $\mathcal{H}$ through o that does not contain η. Thus, $\mathcal{H} \cap \eta$ is a $(k-1)$-dimensional $\mathcal{O}_\eta$-flat, which implies that $\mathcal{O}_\eta$ is nonempty.

Let $\mathcal{O}'$ be the set of $\mathcal{O}$-hyperplanes through o that do not contain η, and $\{\mathcal{H} \cap \eta \mid \mathcal{H} \in \mathcal{O}'\}$ be the set of their intersections with η. All elements of the latter set are $(k-1)$-dimensional $\mathcal{O}_\eta$-planes through o, and their intersection is o. Thus, the intersection of all $(k-1)$-dimensional $\mathcal{O}_\eta$-planes through o is exactly o; that is, $\bigcap \mathcal{O}_\eta = o$. □

4.2 $\mathcal{O}$-Convexity and $\mathcal{O}$-Connectedness

The definition of $\mathcal{O}$-convex sets in multidimensional space is the same as in two dimensions; that is, *a set is $\mathcal{O}$-convex if its intersection with every $\mathcal{O}$-line is connected.* For example, the sets in Fig. 4.3b–e are $\mathcal{O}$-convex for the

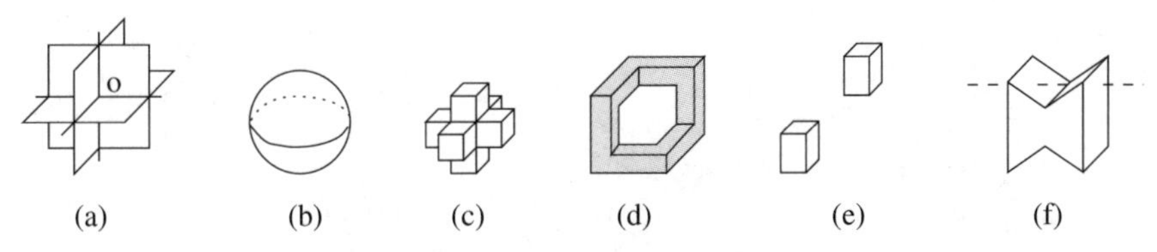

Fig. 4.3. $\mathcal{O}$-convexity in three dimensions

orientation set in Fig. 4.3a. On the other hand, the set in Fig. 4.3f is not $\mathcal{O}$-convex, because its intersection with the dashed $\mathcal{O}$-line is disconnected.

We next consider the properties of planar $\mathcal{O}$-convexity given in Lemma 2.1. Properties 1–5 hold in higher dimensions, and their proofs are the same as in two dimensions; in particular, the intersection of $\mathcal{O}$-convex sets is $\mathcal{O}$-convex. On the other hand, Property 6 does not hold in higher dimensions; that is, a connected $\mathcal{O}$-convex set may not be simply connected. For example, the set in Fig. 4.3d is $\mathcal{O}$-convex with respect to the orthogonal-orientation set, but it is not simply connected.

We now characterize the situations when standard convexity is equivalent to $\mathcal{O}$-convexity. In addition, we determine when different orientation sets give rise to the same $\mathcal{O}$-convex sets.

Lemma 4.5.

1. *$\mathcal{O}$-convexity is equivalent to standard convexity if and only if every line is an $\mathcal{O}$-line.*
2. *$\mathcal{O}_1$-convexity is equivalent to $\mathcal{O}_2$-convexity if and only if the set of all $\mathcal{O}_1$-lines is identical to the set of all $\mathcal{O}_2$-lines.*

Proof. We prove the second statement, which readily implies the first statement. If the set of all $\mathcal{O}_1$-lines is identical to the set of all $\mathcal{O}_2$-lines, the corresponding convexities are equivalent by definition. Suppose, conversely, that some $\mathcal{O}_1$-line is not an $\mathcal{O}_2$-line and consider two distinct points of this $\mathcal{O}_1$-line. The disconnected set formed by these two points is $\mathcal{O}_2$-convex but not $\mathcal{O}_1$-convex, which implies that $\mathcal{O}_1$-convexity is different from $\mathcal{O}_2$-convexity. □

We next characterize $\mathcal{O}$-convex sets in terms of their intersections with $\mathcal{O}$-hyperplanes.

Theorem 4.6. *A set is $\mathcal{O}$-convex if and only if its intersection with every $\mathcal{O}$-hyperplane is $\mathcal{O}$-convex.*

Proof. All $\mathcal{O}$-hyperplanes are convex, which implies that they are $\mathcal{O}$-convex. Since the intersection of two $\mathcal{O}$-convex sets is $\mathcal{O}$-convex, the intersection of an $\mathcal{O}$-convex set with every $\mathcal{O}$-hyperplane is $\mathcal{O}$-convex.

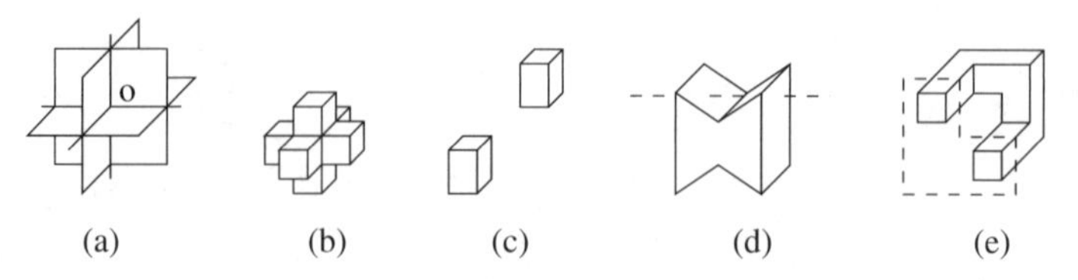

Fig. 4.4. $\mathcal{O}$-connectedness in three dimensions

Suppose, conversely, that the intersection of a set P with every $\mathcal{O}$-hyperplane is $\mathcal{O}$-convex. To show that the intersection of P with every $\mathcal{O}$-line l is connected, we choose some $\mathcal{O}$-hyperplane $\mathcal{H}$ that contains l. Since $P \cap \mathcal{H}$ is $\mathcal{O}$-convex, the intersection of $P \cap \mathcal{H}$ with l is connected. We note that $P \cap \mathcal{H} \cap l = P \cap l$; hence, the intersection of P with l is connected. □

We now describe two special subclasses of $\mathcal{O}$-convex sets, which have a connectedness property; that is, all sets in these subclasses are connected, just like standard convex sets. We define them in terms of their intersections with $\mathcal{O}$-flats. The first subclass is based on the notion of standard connectedness, whereas the second arises from path connectedness.

A set is **path connected** if every two of its points can be connected by a curvilinear segment that is wholly in the set; in particular, the empty set is considered path connected. Note that every path-connected set is connected, whereas some connected sets are not path connected.

Definition 4.3 ($\mathcal{O}$-connectedness). *A set is* **$\mathcal{O}$-connected** *if its intersection with every $\mathcal{O}$-flat is connected. Similarly, a set is* **$\mathcal{O}$-path connected** *if its intersection with every $\mathcal{O}$-flat is path connected.*

For instance, the set in Fig. 4.4b is both $\mathcal{O}$-connected and $\mathcal{O}$-path connected for the orientation set in Fig. 4.4a. On the other hand, the set in Fig. 4.4c is not $\mathcal{O}$-connected because it is disconnected; the set in Fig. 4.4d is not $\mathcal{O}$-connected because its intersection with the dashed $\mathcal{O}$-line is disconnected; and the set in Fig. 4.4e is not $\mathcal{O}$-connected because its intersection with the dashed $\mathcal{O}$-plane is disconnected.

In two dimensions, each connected $\mathcal{O}$-convex set is $\mathcal{O}$-connected, since its intersections with $\mathcal{O}$-lines and with the whole plane are connected; similarly, every path-connected $\mathcal{O}$-convex set in the plane is $\mathcal{O}$-path connected. In higher dimensions, a connected $\mathcal{O}$-convex set may not be $\mathcal{O}$-connected; for example, the set in Fig. 4.4e is path-connected $\mathcal{O}$-convex, but it is not $\mathcal{O}$-connected.

The next result readily follows from the definition of $\mathcal{O}$-connectedness.

Lemma 4.7.

1. *Every translate of an $\mathcal{O}$-connected set is $\mathcal{O}$-connected, and every translate of an $\mathcal{O}$-path-connected set is $\mathcal{O}$-path connected.*

2. *Every convex set is $\mathcal{O}$-path connected, every $\mathcal{O}$-path-connected set is $\mathcal{O}$-connected, and every $\mathcal{O}$-connected set is $\mathcal{O}$-convex.*

Note that the intersection of $\mathcal{O}$-connected sets may not be $\mathcal{O}$-connected; for example, the intersection of the set in Fig. 4.4b with some lines is disconnected, although this set and all lines are $\mathcal{O}$-connected.

We can characterize $\mathcal{O}$-connected sets in terms of their intersections with $\mathcal{O}$-hyperplanes, which is similar to the characterization of $\mathcal{O}$-convex sets in Theorem 4.6.

Theorem 4.8.

1. *A set is $\mathcal{O}$-connected if and only if it is connected and its intersection with every $\mathcal{O}$-hyperplane is $\mathcal{O}$-connected.*
2. *A set is $\mathcal{O}$-path connected if and only if it is path connected and its intersection with every $\mathcal{O}$-hyperplane is $\mathcal{O}$-path connected.*

Proof. We prove only the first statement since the proof of the second statement is similar.

Suppose that P is an $\mathcal{O}$-connected set. We show that P's intersection with every $\mathcal{O}$-hyperplane $\mathcal{H}$ is $\mathcal{O}$-connected by demonstrating that, for every $\mathcal{O}$-flat η, the intersection of $P \cap \mathcal{H}$ with η is connected. Since $\mathcal{H} \cap \eta$ is empty or an $\mathcal{O}$-flat (Lemma 4.3) and P is $\mathcal{O}$-connected, the intersection of $\mathcal{H} \cap \eta$ with P is connected. This intersection is identical to the intersection of $P \cap \mathcal{H}$ with η; hence, the intersection of $P \cap \mathcal{H}$ with η is connected.

Suppose, conversely, that P is a connected set and its intersection with every $\mathcal{O}$-hyperplane is $\mathcal{O}$-connected. To demonstrate that the intersection of P with each $\mathcal{O}$-flat η is connected, we choose some $\mathcal{O}$-hyperplane $\mathcal{H}$ that contains η. Since $P \cap \mathcal{H}$ is $\mathcal{O}$-connected, the intersection of $P \cap \mathcal{H}$ with η is connected. We note that $P \cap \mathcal{H} \cap \eta = P \cap \eta$; hence, the intersection of P with η is connected. □

If the orientation set $\mathcal{O}$ is composed of mutually orthogonal hyperplanes, $\mathcal{O}$-connected sets have one more interesting property; specifically, an orthogonal projection of an $\mathcal{O}$-connected set onto an $\mathcal{O}$-flat is $\mathcal{O}$-connected. We illustrate this property in Fig. 4.5a, where the projection of a three-dimensional $\mathcal{O}$-connected cross is a planar $\mathcal{O}$-connected cross.

Theorem 4.9. *If an orientation set $\mathcal{O}$ consists of mutually orthogonal hyperplanes, then:*

1. *The orthogonal projection of an $\mathcal{O}$-connected set onto an $\mathcal{O}$-flat is $\mathcal{O}$-connected.*
2. *The orthogonal projection of an $\mathcal{O}$-path-connected set onto an $\mathcal{O}$-flat is $\mathcal{O}$-path connected.*

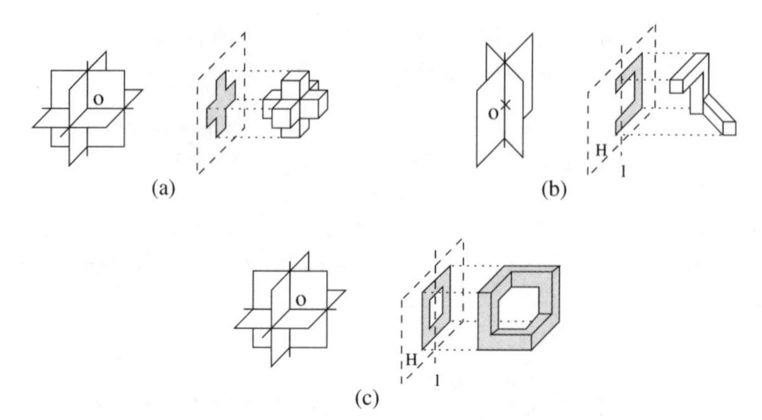

Fig. 4.5. Orthogonal projections of $\mathcal{O}$-connected and $\mathcal{O}$-convex sets onto $\mathcal{O}$-planes

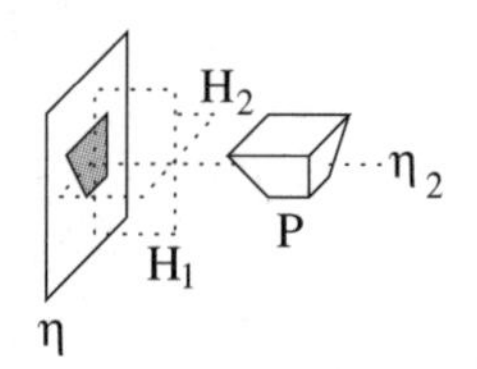

Fig. 4.6. Proof of Theorem 4.9

Proof. We again give a proof for $\mathcal{O}$-connected sets; the same reasoning is applicable to $\mathcal{O}$-path-connected sets. Let P be an $\mathcal{O}$-connected set, let η be an $\mathcal{O}$-flat, and let P_η be the orthogonal projection of P onto η. Note that P_η is connected, because it is a projection of a connected set. We show that, for every $\mathcal{O}$-flat $\eta_1 \neq \mathcal{R}^d$, the intersection of P_η and η_1 is connected.

By definition, the $\mathcal{O}$-flat η_1 is the intersection of several $\mathcal{O}$-hyperplanes, say $\mathcal{H}_1, \mathcal{H}_2, \ldots, \mathcal{H}_n$. If one of these hyperplanes does not intersect η, then the intersection of P_η and η_1 is empty. If all these hyperplanes contain η, then $\eta \subseteq \eta_1$, which implies that the intersection of P_η and η_1 is the set P_η itself, which is connected.

Finally, we consider the case when the hyperplanes $\mathcal{H}_1, \mathcal{H}_2, \ldots, \mathcal{H}_n$ all intersect η and some of them, say $\mathcal{H}_1, \mathcal{H}_2, \ldots, \mathcal{H}_k$, do not contain η. Since the orientation set $\mathcal{O}$ is composed of mutually orthogonal hyperplanes, the $\mathcal{O}$-hyperplanes $\mathcal{H}_1, \mathcal{H}_2, \ldots, \mathcal{H}_k$ are all orthogonal to η, as illustrated in Fig. 4.6. We consider the $\mathcal{O}$-flat $\eta_2 = \mathcal{H}_1 \cap \mathcal{H}_2 \cap \cdots \cap \mathcal{H}_k$; note that $P_\eta \cap \eta_1 = P_\eta \cap \eta_2$, which implies that $P_\eta \cap \eta_1$ is the projection of $P \cap \eta_2$ onto η, as shown in Fig. 4.6. Since P is $\mathcal{O}$-connected, the intersection of P with the $\mathcal{O}$-flat η_2 is connected. Since $P_\eta \cap \eta_1$ is the projection of $P \cap \eta_2$, we conclude that the intersection of P_η and η_1 is also connected. □

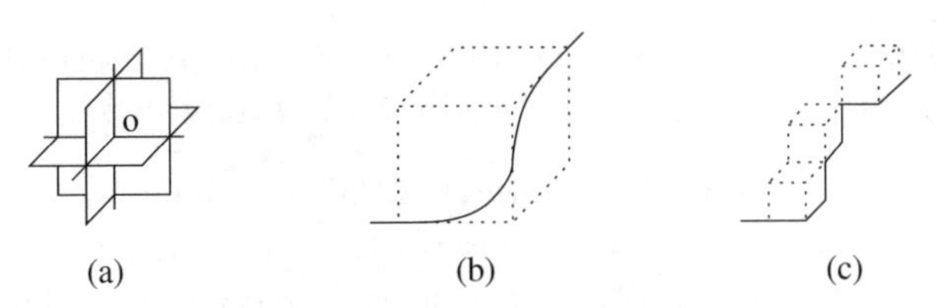

Fig. 4.7. $\mathcal{O}$-connected curves

If elements of the orientation set $\mathcal{O}$ are not mutually orthogonal, the projection of an $\mathcal{O}$-connected object onto an $\mathcal{O}$-flat may not be $\mathcal{O}$-connected. We give an example of such a situation in Fig. 4.5b, where the orientation set contains two planes. The object in Fig. 4.5b is $\mathcal{O}$-connected, but its projection onto the $\mathcal{O}$-plane H is not $\mathcal{O}$-connected, since the intersection of this projection with the dashed $\mathcal{O}$-line l is disconnected.

The orthogonal projection of an $\mathcal{O}$-convex set onto an $\mathcal{O}$-flat may not be $\mathcal{O}$-convex, even for the orthogonal-orientation set. For example, the projection of the $\mathcal{O}$-convex set in Fig. 4.5c onto the $\mathcal{O}$-plane H is not $\mathcal{O}$-convex, since the intersection of this projection with the dashed $\mathcal{O}$-line l is disconnected.

4.3 $\mathcal{O}$-Connected Curves

We next consider $\mathcal{O}$-connected curves and curvilinear segments.

Definition 4.4 (Curves and their segments). *A* **curve** *in d dimensions is the image under a continuous mapping from a line into $\mathcal{R}^d$. A* **segment** *of such a curve is the image of a segment of the line under this mapping.*

We restrict attention to the exploration of simple $\mathcal{O}$-connected curves.

Definition 4.5 (Simple curves). *A curve c is* **simple** *if, for every two points p and q of c, the shortest path from p to q that is wholly in c is a segment of c.*

Informally, the shortest way to reach p from q while remaining in c is to follow c. Self-intersecting curves are not simple, because, if p and q are points on different sides of a loop, the shortest path from p to q does not traverse the loop. Some unusual curves are not simple even though they are not self-intersecting; for example, a Peano curve that covers all the points of a unit square is not simple even though it does not intersect itself.

Note that every line is an $\mathcal{O}$-connected curve. In Fig. 4.7, we show two examples of more complex $\mathcal{O}$-connected curves, which run along the edges of the dotted cubes. We begin by characterizing $\mathcal{O}$-connected curves and curvilinear segments in terms of their intersections with $\mathcal{O}$-hyperplanes.

Lemma 4.10. *A simple curve (curvilinear segment) is $\mathcal{O}$-connected if and only if its intersection with every $\mathcal{O}$-hyperplane is connected.*

Proof. The "only if" part follows directly from the definition of $\mathcal{O}$-connected sets. Suppose, conversely, that the intersection of a simple curve (curvilinear segment) c with every $\mathcal{O}$-hyperplane is connected. We show that c is $\mathcal{O}$-connected by demonstrating that, for every $\mathcal{O}$-flat η and every two points $p, q \in c \cap \eta$, the segment $c[p, q]$ of the curve c is wholly in η, which implies that the intersection of c with every $\mathcal{O}$-flat η is path connected.

If η is an $\mathcal{O}$-flat, then η is the intersection of several $\mathcal{O}$-hyperplanes. For each of these hyperplanes, its intersection with the curve c is connected. Since c is simple, this observation implies that $c[p, q]$ is contained in all of these hyperplanes; thus, $c[p, q]$ is wholly in η. □

Observe that Lemma 4.10 holds only for simple curves. If a curve is not simple, it may not be $\mathcal{O}$-connected even if its intersection with every $\mathcal{O}$-hyperplane is connected. For example, consider a Peano curve in three dimensions that covers all points of a ball's boundary. The intersection of this curve with every plane is connected; however, the curve is not $\mathcal{O}$-connected, since its intersection with any $\mathcal{O}$-line through the ball's center is disconnected.

Lemma 4.11. *A simple curve (curvilinear segment) is $\mathcal{O}$-connected if and only if it is $\mathcal{O}$-path connected.*

Proof. By definition, every $\mathcal{O}$-path-connected curve is $\mathcal{O}$-connected. Suppose, conversely, that a simple curve is $\mathcal{O}$-connected, which means that its intersection with every $\mathcal{O}$-hyperplane is connected. According to the proof of Lemma 4.10, the intersection of this curve with every $\mathcal{O}$-flat is path connected, which implies that the curve is $\mathcal{O}$-path connected. □

Next, we show that every segment of an $\mathcal{O}$-connected curve is $\mathcal{O}$-connected and, conversely, every $\mathcal{O}$-connected curvilinear segment can be extended into an $\mathcal{O}$-connected curve.

Lemma 4.12 (Segment extension).

1. *For each simple $\mathcal{O}$-connected curve c and every two points p and q of c, the segment $c[p, q]$ of the curve c is $\mathcal{O}$-connected.*
2. *For each simple $\mathcal{O}$-connected segment $c[p, q]$, there is a simple $\mathcal{O}$-connected curve c such that $c[p, q]$ is a segment of c.*

Proof.

(1) Let c be an $\mathcal{O}$-connected curve. We have shown in the proof of Lemma 4.10 that, for every $\mathcal{O}$-flat η and every two points $x, y \in c \cap \eta$, the segment of c between x and y is contained in η; in particular, this observation

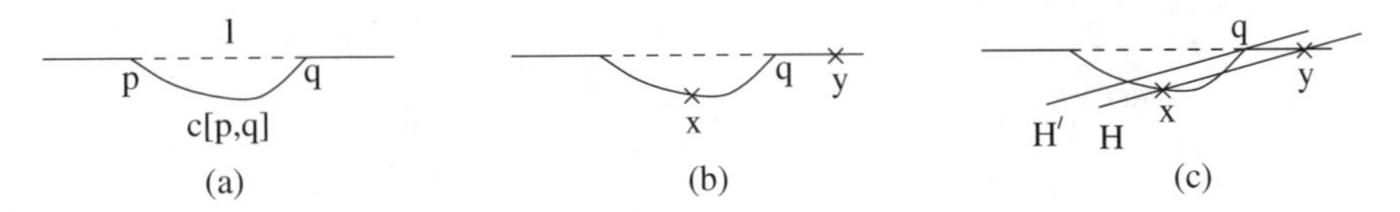

Fig. 4.8. Proof of Lemma 4.12

holds for every two points $x, y \in c[p,q] \cap \eta$. Therefore, the intersection of $c[p,q]$ with every $\mathcal{O}$-flat η is connected, which means that $c[p,q]$ is $\mathcal{O}$-connected.

(2) Let $c[p,q]$ be an $\mathcal{O}$-connected curvilinear segment, and l be the line through its endpoints p and q, as shown in Fig. 4.8a. We consider the curve c obtained from l by replacing the line segment between p and q with the curvilinear segment $c[p,q]$; this curve is shown by solid lines in Fig. 4.8a.

We prove that c is $\mathcal{O}$-connected by showing that, for every $\mathcal{O}$-hyperplane $\mathcal{H}$ and every two points $x, y \in c \cap \mathcal{H}$, the segment of c between x and y is wholly in $\mathcal{H}$. This proof implies that the intersection of c with every $\mathcal{O}$-hyperplane is connected; hence, c is $\mathcal{O}$-connected by Lemma 4.10.

If x and y are in $c[p,q]$, then the segment of $c[p,q]$ between x and y is wholly in $\mathcal{H}$, because $c[p,q] \cap \mathcal{H}$ is connected and $c[p,q]$ is simple. If x and y are in l, then $l \subseteq \mathcal{H}$, which implies that p and q are in $\mathcal{H}$. Since the intersection of $c[p,q]$ and $\mathcal{H}$ is connected and $c[p,q]$ is simple, we conclude that $c[p,q]$ is wholly in $\mathcal{H}$. Thus, the curve c is contained in $\mathcal{H}$, which implies that the segment of c between x and y is in $\mathcal{H}$.

Finally, we consider the case when x is in $c[p,q]$ and y is in l; that is, y is in one of the two rays extending $c[p,q]$, as shown in Fig. 4.8b. Without loss of generality, we assume that q (rather than p) is between x and y on the curve c, as shown in Fig. 4.8b. If $q \notin \mathcal{H}$, then the intersection of the segment $c[p,q]$ and the $\mathcal{O}$-hyperplane $\mathcal{H}'$ through q parallel to $\mathcal{H}$ is disconnected, as shown in Fig. 4.8c, contradicting the $\mathcal{O}$-connectedness of $c[p,q]$. We conclude that $q \in \mathcal{H}$; therefore, the line segment joining q and y is in $\mathcal{H}$, and the segment of $c[p,q]$ between x and q is also in $\mathcal{H}$. Thus, the segment of c between x and y is wholly in $\mathcal{H}$. □

If we cut a segment from an $\mathcal{O}$-connected curve and replace it with another $\mathcal{O}$-connected segment, as shown in Fig. 4.9a, then the resulting curve is also $\mathcal{O}$-connected.

Lemma 4.13 (Cutting and pasting). *Let p and q be two points of a simple $\mathcal{O}$-connected curve c. If we replace the part of c between p and q with another simple $\mathcal{O}$-connected segment, then the resulting curve c' is also $\mathcal{O}$-connected.*

Proof. We show that the curve c' is $\mathcal{O}$-connected by demonstrating that, for every $\mathcal{O}$-hyperplane $\mathcal{H}$ and every two points $x, y \in c' \cap \mathcal{H}$, the segment of c' between x and y is wholly in $\mathcal{H}$. Then, the intersection of c' with every $\mathcal{O}$-hyperplane $\mathcal{H}$ is connected, which implies that c' is $\mathcal{O}$-connected by

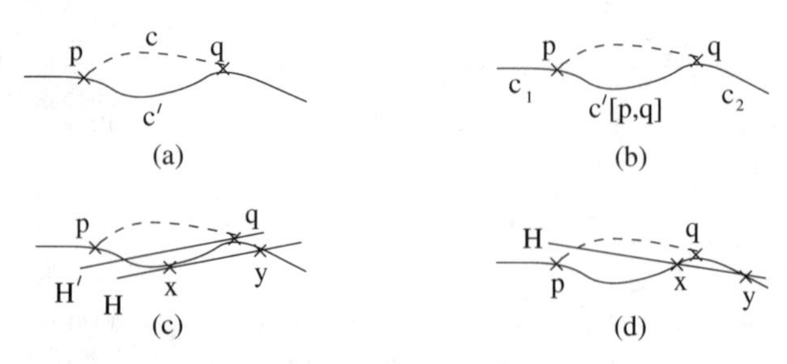

Fig. 4.9. Proof of Lemma 4.13

Lemma 4.10. We denote the three parts of c' by c_1, $c'[p, q]$ and c_2, as shown in Fig. 4.9b. Note that c_1 and c_2 are $\mathcal{O}$-connected by Lemma 4.12, because they are parts of the $\mathcal{O}$-connected curve c.

If x and y are in $c'[p, q]$, then the segment of $c'[p, q]$ between x and y is wholly in $\mathcal{H}$, because $c'[p, q] \cap \mathcal{H}$ is connected and $c'[p, q]$ is simple. Similarly, if $x, y \in c_1$ (or $x, y \in c_2$), then the segment of the curve between x and y is wholly in $\mathcal{H}$, because $c_1 \cap \mathcal{H}$ is connected.

If $x \in c_1$ and $y \in c_2$, then the segment of the $\mathcal{O}$-connected curve c between x and y is contained in $\mathcal{H}$, which implies that p and q are in $\mathcal{H}$. Since the segment $c'[p, q]$ is $\mathcal{O}$-connected and its endpoints p and q are in $\mathcal{H}$, we conclude that $c'[p, q]$ is in $\mathcal{H}$; thus, the segment of c' between x and y is wholly in $\mathcal{H}$.

Finally, consider the case when $x \in c'[p, q]$ and $y \in c_2$ (or $y \in c_1$), as shown in Fig. 4.9c. We first show, by contradiction, that $q \in \mathcal{H}$. Suppose that $q \notin \mathcal{H}$. If p and q are "on the same side" of $\mathcal{H}$, then the intersection of $c'[p, q]$ and the $\mathcal{O}$-hyperplane $\mathcal{H}'$ through q (or through p) parallel to $\mathcal{H}$ is disconnected, as shown in Fig. 4.9c, contradicting the $\mathcal{O}$-connectedness of $c'[p, q]$. If $p \in \mathcal{H}$, or p and q are "on different sides" of $\mathcal{H}$, as shown in Fig. 4.9d, then the intersection of the curve c and the $\mathcal{O}$-hyperplane $\mathcal{H}$ is disconnected, since c contains the points p, q and y (in this order), which again contradicts the $\mathcal{O}$-connectedness of c. We conclude that $q \in \mathcal{H}$; therefore, the segment of $c'[p, q]$ between x and q and the segment of c_2 between q and y are both in $\mathcal{H}$. Thus, the segment of c' between x and y is wholly in $\mathcal{H}$. □

Finally, we state a condition under which the catenation of several curvilinear segments is an $\mathcal{O}$-connected segment.

Lemma 4.14 (Catenation). *Let $p_0, p_1, \ldots, p_n$ be a sequence of points connected by simple curvilinear segments $c[p_0, p_1], c[p_1, p_2], \ldots, c[p_{n-1}, p_n]$. The union of these segments is an $\mathcal{O}$-connected segment if and only if the following two conditions hold:*

1. *Each of the segments is $\mathcal{O}$-connected.*
2. *For every $\mathcal{O}$-hyperplane $\mathcal{H}$, if $\mathcal{H}$ intersects two segments $c[p_{k-1}, p_k]$ and $c[p_m, p_{m+1}]$, where $k \leq m$, then the points $p_k, p_{k+1}, \ldots, p_m$ are in $\mathcal{H}$.*

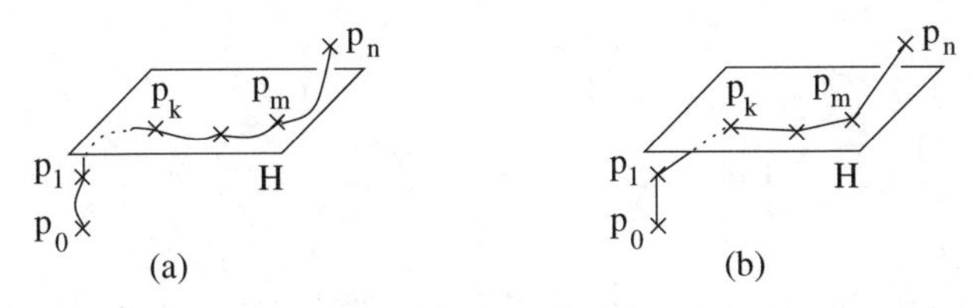

Fig. 4.10. Proof of Lemma 4.14

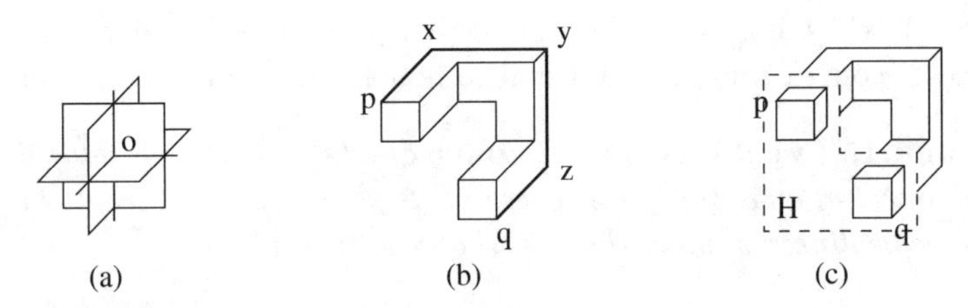

Fig. 4.11. Generalized visibility in three dimensions

Proof. The union of the segments $c[p_0, p_1], c[p_1, p_2], \dots, c[p_{n-1}, p_n]$ is a curvilinear segment, denoted by $c[p_0, p_n]$. Suppose that this segment is $\mathcal{O}$-connected. Then, every segment $c[p_k, p_{k+1}]$ of $c[p_0, p_n]$ is $\mathcal{O}$-connected by Lemma 4.12. Furthermore, the intersection of $c[p_0, p_n]$ with every $\mathcal{O}$-hyperplane $\mathcal{H}$ is connected. Hence, if $\mathcal{H}$ intersects segments $c[p_{k-1}, p_k]$ and $c[p_m, p_{m+1}]$, then the points $p_k, \dots, p_m$ are in $\mathcal{H}$, as illustrated in Fig. 4.10a.

To prove the converse, suppose that the segments $c[p_0, p_1], c[p_1, p_2], \dots,$ $c[p_{n-1}, p_n]$ are line segments, as shown in Fig. 4.10b. Then, Condition 2 immediately implies that the intersection of the polygonal line $c[p_0, p_n]$ with every $\mathcal{O}$-hyperplane is connected; hence, $c[p_0, p_n]$ is $\mathcal{O}$-connected by Lemma 4.10. If we now replace every line segment $c[p_k, p_{k+1}]$ with an arbitrary $\mathcal{O}$-connected segment, the resulting new segment $c[p_0, p_n]$ is also $\mathcal{O}$-connected by Lemma 4.13. □

4.4 Visibility

In standard convexity, two points of a set are *visible* to each other if the line segment joining them is wholly in the set. For example, the points p and x in Fig. 4.11b are visible to each other, whereas p and q are not. Clearly, a set is standard convex if and only if every two of its points are visible to each other.

Since $\mathcal{O}$-convexity is weaker than standard convexity, $\mathcal{O}$-convex sets may not satisfy this visibility condition. For instance, the set in Fig. 4.11b is $\mathcal{O}$-convex for the orientation set in Fig. 4.11a, and its points p and q are not visible to each other.

We define a weaker visibility, which enables us to characterize $\mathcal{O}$-convex sets; specifically, two points are considered visible to each other if there is

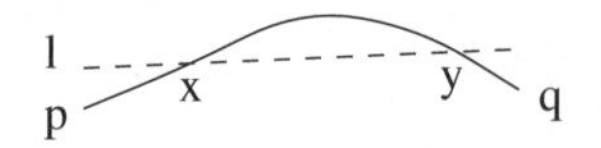

Fig. 4.12. Proof of Theorem 4.15

an $\mathcal{O}$-convex curvilinear segment joining them that is wholly in the set. For example, we can join p and q in Fig. 4.11b with the $\mathcal{O}$-convex polygonal line (p, x, y, z, q), which is contained in the set. We give a visibility characterization for closed $\mathcal{O}$-convex sets; its extension to nonclosed sets is an open problem.

Theorem 4.15 (Visibility for $\mathcal{O}$-convex sets). *A closed path-connected set is $\mathcal{O}$-convex if and only if every two of its points can be joined by a simple $\mathcal{O}$-convex curvilinear segment that is wholly in the set.*

Proof. Suppose that every two points of a set P can be joined by a simple $\mathcal{O}$-convex segment contained in P. Observe that, if a line through two points of P is an $\mathcal{O}$-line, then the only simple $\mathcal{O}$-convex curvilinear segment joining these points is a line segment; therefore, the line segment joining them is in P. This observation implies that the intersection of every $\mathcal{O}$-line with P is connected, which means that P is $\mathcal{O}$-convex.

Suppose, conversely, that a closed set P is path connected and $\mathcal{O}$-convex. To demonstrate that every two points p and q of P can be connected by a simple $\mathcal{O}$-convex segment, we consider a shortest simple curvilinear segment $c[p, q]$ joining p and q in P. Note that such a shortest segment exists because P is closed. We prove, by contradiction, that this segment is $\mathcal{O}$-convex.

If $c[p, q]$ is not $\mathcal{O}$-convex, then its intersection with some $\mathcal{O}$-line l is disconnected, as shown in Fig. 4.12. Therefore, there are two points $x, y \in c[p, q] \cap l$ such that the segment of the line l between x and y is not in $c[p, q]$. On the other hand, since P is $\mathcal{O}$-convex, this segment is wholly in P. If we replace the segment of $c[p, q]$ between x and y with the line segment joining x and y, then we obtain a shorter path from p to q in P, which contradicts the assumption that $c[p, q]$ is a shortest path. □

We can characterize $\mathcal{O}$-path-connected sets in a similar way, through $\mathcal{O}$-connected segments joining their points. This type of visibility is stronger than $\mathcal{O}$-convex visibility; that is, two points sometimes cannot be joined by an $\mathcal{O}$-connected segment even when they can be joined by an $\mathcal{O}$-convex segment. For example, there is no $\mathcal{O}$-connected path from p to q in Fig. 4.11c, because the intersection of the $\mathcal{O}$-plane H with every path between p and q is disconnected.

Theorem 4.16 (Visibility for $\mathcal{O}$-path-connected sets). *A closed set is $\mathcal{O}$-path connected if and only if every two of its points can be joined by a simple $\mathcal{O}$-connected curvilinear segment that is wholly in the set.*

Proof. Suppose that every two points of a set P can be joined by a simple $\mathcal{O}$-connected segment. If two such points are in some $\mathcal{O}$-flat, then the simple $\mathcal{O}$-connected segment joining them is wholly in this $\mathcal{O}$-flat by the proof of Lemma 4.10. Therefore, the intersection of P with every $\mathcal{O}$-flat is path connected, which means that P is $\mathcal{O}$-path connected.

We use induction on the dimension d to prove that, conversely, every two points of an $\mathcal{O}$-path-connected set P can be joined by a simple $\mathcal{O}$-connected path in P. In two dimensions, every two points of an $\mathcal{O}$-connected set can be joined by a simple $\mathcal{O}$-convex path according to Theorem 4.15, and every $\mathcal{O}$-convex path is $\mathcal{O}$-connected, which establishes the induction basis.

The proof of the induction step consists of three parts. First, we show that, if two points p and q of an $\mathcal{O}$-connected set P in d dimensions are contained in some $\mathcal{O}$-hyperplane, then they can be joined by a simple $\mathcal{O}$-connected path in P. Second, we consider the case when there is no $\mathcal{O}$-hyperplane through p and q; we define the $\mathcal{O}$-block of p and q, and show that there is a path from p to q contained in the intersection of this $\mathcal{O}$-block with P. Third, we use this result to construct a simple $\mathcal{O}$-connected path from p to q.

Suppose that points p and q of P belong to some $\mathcal{O}$-hyperplane $\mathcal{H}$. Recall that we may view $\mathcal{H}$ as an independent $(d-1)$-dimensional space and define the corresponding orientation set $\mathcal{O}_{\mathcal{H}}$ formed by $(d-2)$-dimensional $\mathcal{O}$-flats. This orientation set gives rise to $\mathcal{O}_{\mathcal{H}}$-path-connected sets, whose intersections with $\mathcal{O}_{\mathcal{H}}$-flats are path connected. Since $\mathcal{O}_{\mathcal{H}}$-flats are $\mathcal{O}$-flats, a set in $\mathcal{H}$ is $\mathcal{O}_{\mathcal{H}}$-path connected if and only if it is $\mathcal{O}$-path connected. The intersection of P with $\mathcal{H}$ is $\mathcal{O}$-path connected by Theorem 4.8, which implies that it is $\mathcal{O}_{\mathcal{H}}$-path connected. By the induction hypothesis, p and q can be joined by a simple $\mathcal{O}_{\mathcal{H}}$-connected segment in $P \cap \mathcal{H}$; therefore, they can be joined by a simple $\mathcal{O}$-connected segment in P.

Now suppose that there is no $\mathcal{O}$-hyperplane through p and q. We define the d-dimensional $\mathcal{O}$-block of p and q, and show that there is a path from p to q in the intersection of this $\mathcal{O}$-block with P. Let S_p be the intersection of all the halfspaces, containing q, whose boundaries are $\mathcal{O}$-hyperplanes through p; note that S_p is a polyhedral angle with vertex p, as shown in Fig. 4.13a. Similarly, let S_q be the intersection of all the halfspaces, containing p, whose boundaries are $\mathcal{O}$-hyperplanes through q, as shown in Fig. 4.13b. The $\mathcal{O}$-block of p and q, denoted by $\mathcal{O}$-block(p, q), is the intersection of S_p and S_q, as shown in Fig. 4.13c. Observe that every $\mathcal{O}$-connected path from p to q is wholly in $\mathcal{O}$-block(p, q), because the intersection of such a path with every $\mathcal{O}$-hyperplane is connected.

To show that there is a path from p to q in $P \cap \mathcal{O}$-block(p, q), we consider some path $c[p, q] \subseteq P$ from p to q. If $c[p, q]$ is not wholly in S_p, we select a point x in the intersection of $c[p, q]$ with the boundary of S_p such that the segment of $c[p, q]$ between x and q is wholly in S_p, as illustrated by Fig. 4.13d. Note that there is some $\mathcal{O}$-hyperplane $\mathcal{H}$ through p and x; thus, we can join

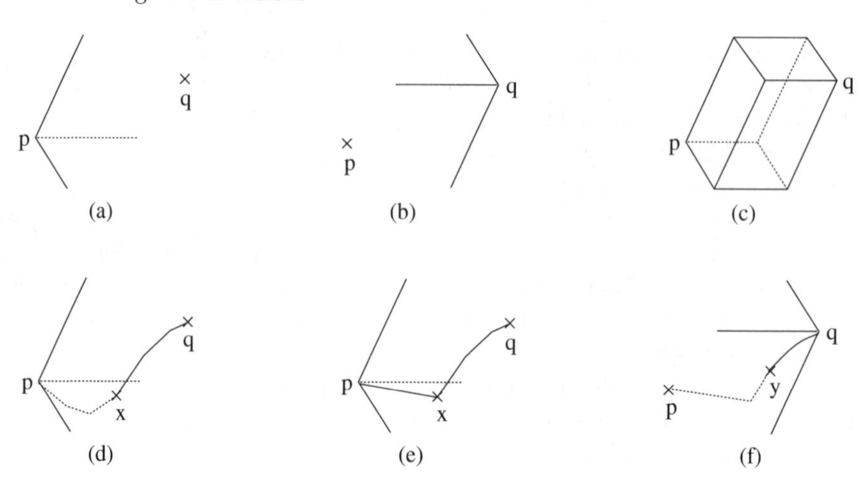

Fig. 4.13. Proof of Theorem 4.16

p and x by an $\mathcal{O}$-connected path in $P \cap \mathcal{H}$. We replace the segment of $c[p, q]$ between p and x with this $\mathcal{O}$-connected path, thus obtaining a new path $c_1[p, q]$ from p to q, which is contained in $P \cap S_p$, as shown in Fig. 4.13e.

If $c_1[p, q]$ is not wholly in S_q, we select a point y in the intersection of $c_1[p, q]$ with the boundary of S_q such that the segment of $c_1[p, q]$ between p and y is wholly in S_q, as shown in Fig. 4.13f. We replace the segment of $c_1[p, q]$ between y and q with an $\mathcal{O}$-connected path in P, thus obtaining a new path from p to q, which is wholly in $P \cap \mathcal{O}$-block(p, q).

Finally, we construct a simple $\mathcal{O}$-connected path in P from p to q. We consider some path from p to q in $P \cap \mathcal{O}$-block(p, q) and choose a point z in this path such that the distance between p and z is equal to the distance between z and q. We next consider some path from p to z in $P \cap \mathcal{O}$-block(p, z) and some path from z to q in $P \cap \mathcal{O}$-block(z, q), and choose a point in each of these two paths, in the same way as we have chosen the point z in the path from p to q.

We recursively repeat this point-selection operation; on the nth level of recursion, we obtain $2^n - 1$ intermediate points, connected by 2^n segments that form a path from p to q. The distances between consecutive points converge to zero as n tends to infinity. By construction, the intersection of this path with every $\mathcal{O}$-hyperplane through any of the intermediate points is connected.

The closure of the set of points selected in infinitely many recursive steps is a simple path from p to q. Furthermore, since P is closed, this path is contained in P. The intersection of the constructed path with every $\mathcal{O}$-hyperplane is connected, which implies that the path is $\mathcal{O}$-connected by Lemma 4.10. □

Summary

We have generalized $\mathcal{O}$-convexity to higher dimensions and demonstrated that the properties of $\mathcal{O}$-convex sets are similar to the properties of standard convex sets. The main property of standard convexity that does not generalize to $\mathcal{O}$-convexity is connectedness; that is, an $\mathcal{O}$-convex set may be disconnected. To bridge this difference, we have introduced $\mathcal{O}$-connected sets, which are always connected, and demonstrated that their properties are also similar to those of standard convex sets. The main results include the characterization of $\mathcal{O}$-convex and $\mathcal{O}$-connected sets in terms of their intersections with $\mathcal{O}$-hyperplanes (Theorems 4.6 and 4.8), properties of $\mathcal{O}$-connected segments and curves (Lemmas 4.12–4.14), and visibility results for closed sets (Theorems 4.15 and 4.16).

5

Generalized Halfspaces

We introduce $\mathcal{O}$-halfspaces and explore their relationship to $\mathcal{O}$-convex and $\mathcal{O}$-connected sets. First, we give basic properties of $\mathcal{O}$-halfspaces and compare them with standard halfspaces (Sect. 5.1). Then, we define directed $\mathcal{O}$-halfspaces, which are a subclass of $\mathcal{O}$-halfspaces with several special properties (Sect. 5.2). Finally, we characterize $\mathcal{O}$-halfspaces in terms of their boundaries (Sect. 5.3) and complements (Sect. 5.4).

5.1 $\mathcal{O}$-Halfspaces

The notion of $\mathcal{O}$-halfspaces is a higher-dimensional analog of $\mathcal{O}$-halfplanes, described in Sect. 2.2.

Definition 5.1 ($\mathcal{O}$-halfspaces). *An* **$\mathcal{O}$-halfspace** *is a closed set whose intersection with every $\mathcal{O}$-line is empty, a ray or a line.*

The sets in Fig. 5.1b–e are $\mathcal{O}$-halfspaces for the orientation set in Fig. 5.1a. We use dotted lines to show infinite planar regions at the boundaries of these $\mathcal{O}$-halfspaces. Unlike standard halfspaces, $\mathcal{O}$-halfspaces may be disconnected; for instance, the $\mathcal{O}$-halfspace in Fig. 5.1e consists of three components located around the dashed cube.

Observe that Properties 1–3 of $\mathcal{O}$-halfplanes, given in Lemma 2.5, readily generalize to higher dimensions. Specifically, every translate of an $\mathcal{O}$-halfspace is an $\mathcal{O}$-halfspace, every standard halfspace is an $\mathcal{O}$-halfspace, and every $\mathcal{O}$-halfspace is $\mathcal{O}$-convex.

Lemma 5.1. *An $\mathcal{O}$-convex set P is an $\mathcal{O}$-halfspace if and only if, for every point p in P and every $\mathcal{O}$-line l, one of the two parallel-to-l rays with endpoint p is wholly in P.*

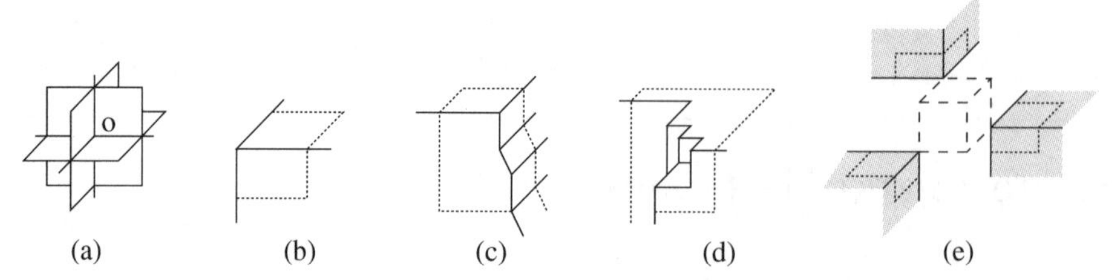

Fig. 5.1. $\mathcal{O}$-halfspaces in three dimensions

Proof. Clearly, this ray condition holds for every $\mathcal{O}$-halfspace; we need to show that, conversely, if P satisfies the ray condition, then P is an $\mathcal{O}$-halfspace. Since P is $\mathcal{O}$-convex, its intersection with every $\mathcal{O}$-line is empty, a point, a line segment, a ray or a line. The ray condition implies that P's intersection with an $\mathcal{O}$-line is neither a point nor a line segment, which means that this intersection is empty, a ray or a line. □

We next characterize disconnected $\mathcal{O}$-halfspaces in terms of their connected components, and give a bound on the number of their components.

Theorem 5.2.

1. *A disconnected set is an $\mathcal{O}$-halfspace if and only if every connected component of the set is an $\mathcal{O}$-halfspace and no $\mathcal{O}$-line intersects two components.*
2. *In d dimensions, if the orientation set $\mathcal{O}$ has the point-intersection property, then the number of components of an $\mathcal{O}$-halfspace is at most 2^{d-1}.*

Proof.

(1) Clearly, if P is the union of $\mathcal{O}$-halfspaces and no $\mathcal{O}$-line intersects two of them, then P is an $\mathcal{O}$-halfspace. If some connected component of P is not an $\mathcal{O}$-halfspace, then the intersection of this component with some $\mathcal{O}$-line is nonempty, not a ray and not a line; therefore, the intersection of P with this $\mathcal{O}$-line is nonempty, not a ray and not a line. Finally, if some $\mathcal{O}$-line intersects two components, then the intersection of P with this $\mathcal{O}$-line is disconnected.

(2) We use induction on the dimension d to show that, for any $2^{d-1}+1$ points of an $\mathcal{O}$-halfspace P, two of them are in the same connected component. We have proved this result for $\mathcal{O}$-halfplanes in Lemma 2.8, which provides an induction basis. The induction step is based on Lemma 5.4 (page 57), where we show that *the intersection of an $\mathcal{O}$-halfspace P with an $\mathcal{O}$-hyperplane $\mathcal{H}$ is an $\mathcal{O}$-halfspace in the $(d-1)$-dimensional space $\mathcal{H}$;* that is, the intersection of $P \cap \mathcal{H}$ with every $\mathcal{O}$-line contained in $\mathcal{H}$ is empty, a ray or a line.

We denote the $2^{d-1}+1$ points in P by $p_0, p_1, \ldots, p_{2^{d-1}}$; in Fig. 5.2, we illustrate the proof for $d = 3$. By Lemma 4.2, we can choose some $\mathcal{O}$-line l and an $\mathcal{O}$-hyperplane $\mathcal{H}$ that intersects l and does not contain it, as shown in Fig. 5.2a. For every point p_k, one of the two parallel-to-l rays with endpoint p_k

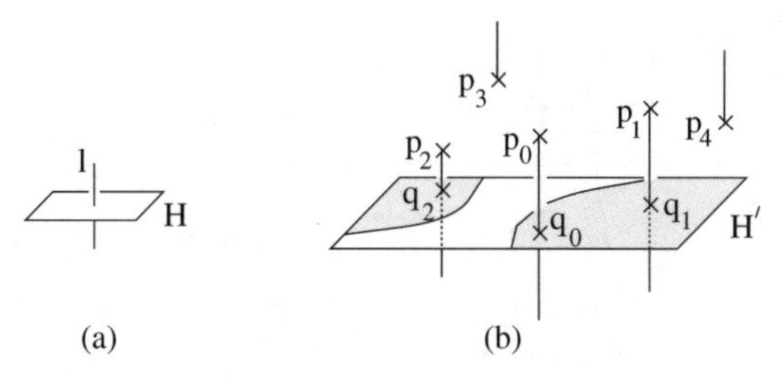

Fig. 5.2. Proof of Theorem 5.2

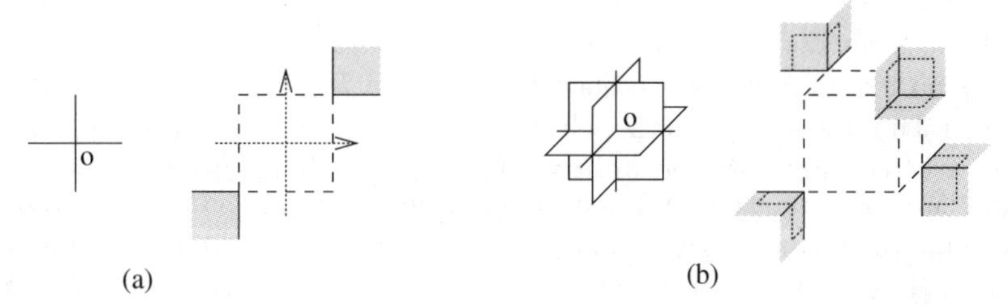

Fig. 5.3. $\mathcal{O}$-halfspaces with 2^{d-1} components in **(a)** two and **(b)** three dimensions

is contained in P. Thus, there are $2^{d-1}+1$ parallel rays in P, and at least $2^{d-2}+1$ of them have the same direction; we assume that the endpoints of these rays are $p_0, p_1, \ldots, p_{2^{d-2}}$. We select an $\mathcal{O}$-hyperplane $\mathcal{H}'$, parallel to $\mathcal{H}$, that intersects these $2^{d-2}+1$ same-direction rays, and denote the respective points of their intersection with $\mathcal{H}'$ by $q_0, q_1, \ldots, q_{2^{d-2}}$.

The intersection of P and $\mathcal{H}'$, shaded in Fig. 5.2b, is an $\mathcal{O}$-halfspace in the $(d-1)$-dimensional space $\mathcal{H}$. By the induction hypothesis, some points q_k and q_m belong to the same connected component of $P \cap \mathcal{H}'$, which implies that p_k and p_m belong to the same component of P. □

If the orientation set $\mathcal{O}$ does not have the point-intersection property, an $\mathcal{O}$-halfspace may have infinitely many components. For example, if there is only one $\mathcal{O}$-line through o, then any collection of lines parallel to this $\mathcal{O}$-line forms an $\mathcal{O}$-halfspace. We next show that the upper bound on the number of components given in Theorem 5.2 is tight.

Example: $\mathcal{O}$-halfspace with 2^{d-1} connected components.
Consider the orthogonal-orientation set in d dimensions, which comprises d mutually orthogonal hyperplanes; note that it has the point-intersection property. We construct an $\mathcal{O}$-halfspace P whose components are rectangular polyhedral angles, that is, regions of space bounded by d mutually orthogonal hyperplanes through a common point. We illustrate this construction in Fig. 5.3, where the polyhedral angles are shaded.

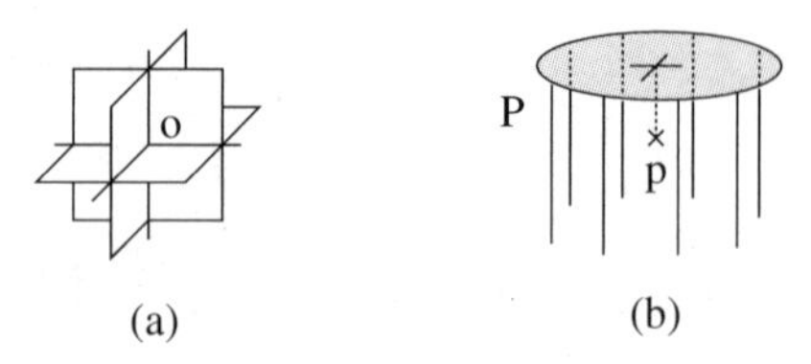

Fig. 5.4. $\mathcal{O}$-convex set P that is not the intersection of $\mathcal{O}$-halfspaces, as every $\mathcal{O}$-halfspace containing P also contains the point $p \notin P$

We choose a cube, shown by dashed lines in Fig. 5.3, whose facets are parallel to elements of $\mathcal{O}$; note that this cube has 2^d vertices. We select 2^{d-1} vertices no two of which are adjacent and define P as the union of the 2^{d-1} polyhedral angles vertical to the selected angles of the cube, that is, symmetric to the selected angles with respect to the corresponding vertices.

We can describe P analytically using $\mathcal{O}$-lines through the cube's center as coordinate axes, and letting one of the cube's vertices be $(1, 1, \ldots, 1)$. The equation for P is

$$|x_1|, |x_2|, \ldots, |x_d| \geq 1 \text{ and } x_1 \cdot x_2 \cdot \cdots \cdot x_d > 0.$$

To show that P is indeed an $\mathcal{O}$-halfspace, we observe that any given $\mathcal{O}$-line either intersects one component, in which case the intersection is a ray, or does not intersect any components.

In two dimensions, if an $\mathcal{O}$-convex set is closed and connected, then it is formed by the intersection of $\mathcal{O}$-halfplanes, as shown in Lemma 2.6; however, this result does not generalize to higher dimensions.

Example: $\mathcal{O}$-convex set that is not the intersection of $\mathcal{O}$-halfspaces. Consider the orthogonal-orientation set in Fig. 5.4a, and the $\mathcal{O}$-convex set P in Fig. 5.4b, which is a horizontal disk on eight vertical, equally spaced "pillars." These pillars are rays located in such a way that no $\mathcal{O}$-line intersects two of them. Every $\mathcal{O}$-halfspace containing P also contains the point $p \notin P$, located under the center of the disk, which means that the intersection of all $\mathcal{O}$-halfspaces containing P is a proper superset of P.

5.2 Directed $\mathcal{O}$-Halfspaces

We next consider directed $\mathcal{O}$-halfspaces, which are a higher-dimensional analog of directed $\mathcal{O}$-halfplanes, described in Sect. 2.2. We define directed $\mathcal{O}$-halfspaces through the notion of ray direction; recall that two rays have the same direction if they are translates of each other.

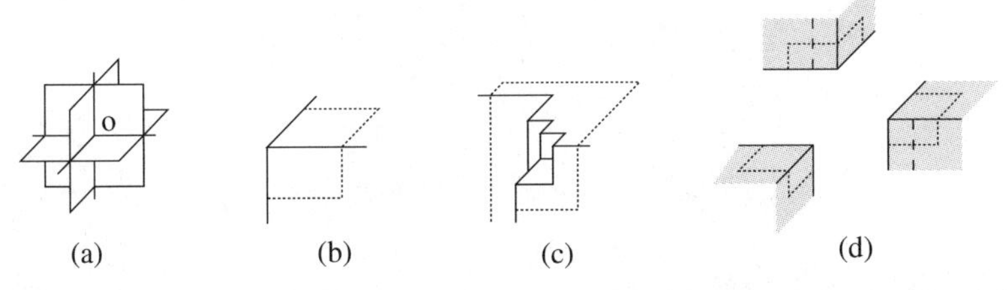

Fig. 5.5. Directed $\mathcal{O}$-halfspaces **(b,c)** and a nondirected $\mathcal{O}$-halfspace **(d)**

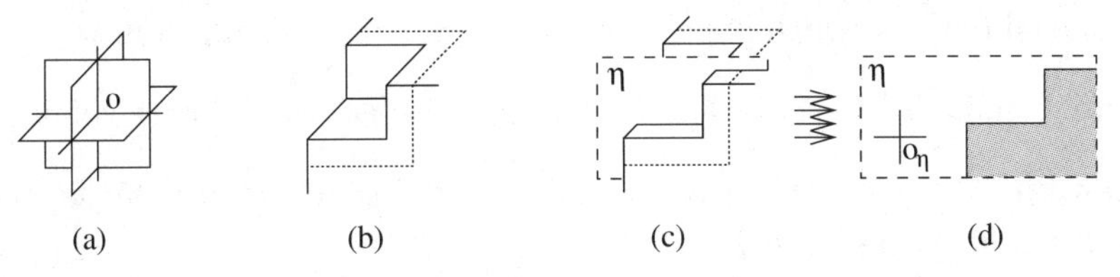

Fig. 5.6. Illustration of Lemma 5.4, which states that, for any orientation set **(a)**, the intersection of an $\mathcal{O}$-halfspace **(b)** with an $\mathcal{O}$-flat η **(c)** is an $\mathcal{O}_\eta$-halfspace **(d)**

Definition 5.2 (Directed $\mathcal{O}$-halfspaces). *An $\mathcal{O}$-halfspace is* **directed** *if, for every two parallel $\mathcal{O}$-lines whose intersection with the $\mathcal{O}$-halfspace yields rays, these rays have the same direction (rather than opposite directions).*

The $\mathcal{O}$-halfspaces in Fig. 5.5b,c are directed for the orthogonal-orientation set in Fig. 5.5a. On the other hand, the $\mathcal{O}$-halfspace in Fig. 5.5d is not directed, because the two dashed rays, formed by its intersection with vertical $\mathcal{O}$-lines, have opposite directions.

The next result readily follows from the definition of directed $\mathcal{O}$-halfspaces.

Lemma 5.3.

1. *Every translate of a directed $\mathcal{O}$-halfspace is a directed $\mathcal{O}$-halfspace.*
2. *Every standard $\mathcal{O}$-halfspace is a directed $\mathcal{O}$-halfspace.*

We now consider the intersection of a directed $\mathcal{O}$-halfspace with an arbitrary $\mathcal{O}$-flat η. The orientation set $\mathcal{O}_\eta$, which is formed by the intersection of η with $\mathcal{O}$-hyperplanes, as described in Sect. 4.1, gives rise to lower-dimensional $\mathcal{O}_\eta$-halfspaces in the space η. These lower-dimensional $\mathcal{O}_\eta$-halfspaces are defined in the same way as $\mathcal{O}$-halfspaces, in terms of their intersection with $\mathcal{O}_\eta$-lines. The following result, illustrated in Fig. 5.6, allows the use of induction on the dimension d in the exploration of $\mathcal{O}$-halfspaces.

Lemma 5.4. *The intersection of a (directed) $\mathcal{O}$-halfspace with an $\mathcal{O}$-flat η is a (directed) $\mathcal{O}_\eta$-halfspace.*

Proof. Suppose that P is an $\mathcal{O}$-halfspace. For every $\mathcal{O}$-line $l \subseteq \eta$, we have $l \cap (P \cap \eta) = l \cap P$, which implies that the intersection of l with $P \cap \eta$ is empty,

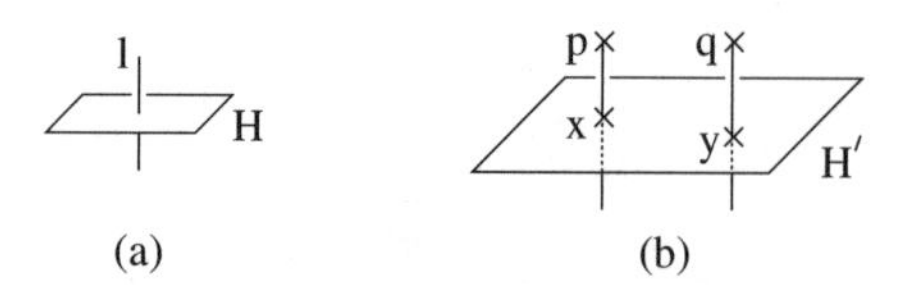

Fig. 5.7. Proof of Theorem 5.5

a ray or a line; hence, $P \cap \eta$ is an $\mathcal{O}_\eta$-halfspace. If P is a directed $\mathcal{O}$-halfspace, the rays formed by the intersection of $P \cap \eta$ with parallel $\mathcal{O}$-lines in η have the same direction, which implies that $P \cap \eta$ is a directed $\mathcal{O}_\eta$-halfspace. □

We use Lemma 5.4 to show that directed $\mathcal{O}$-halfspaces are $\mathcal{O}$-path connected.

Theorem 5.5. *If an orientation set $\mathcal{O}$ has the point-intersection property, then every directed $\mathcal{O}$-halfspace is $\mathcal{O}$-path connected.*

Proof. We first prove, by induction on the dimension d, that all directed $\mathcal{O}$-halfspaces are path connected. We have already proved this result for directed $\mathcal{O}$-halfplanes in Lemma 2.8, which provides an induction basis. We now consider a directed $\mathcal{O}$-halfspace P in d dimensions, and show that every two points $p, q \in P$ can be joined by a path in P. By Lemma 4.2, we can select some $\mathcal{O}$-line l and an $\mathcal{O}$-hyperplane $\mathcal{H}$ that intersects l and does not contain it, as shown in Fig. 5.7a. Since P is a directed $\mathcal{O}$-halfspace, there are two rays parallel to l, with endpoints p and q, that are contained in P and have the same direction. We choose an $\mathcal{O}$-plane $\mathcal{H}'$, parallel to $\mathcal{H}$, that intersects these two rays, and denote the respective intersection points by x and y, as shown in Fig. 5.7b. Note that $P \cap \mathcal{H}'$ is a directed $\mathcal{O}_{\mathcal{H}'}$-halfspace by Lemma 5.4, and the lower-dimensional orientation set $\mathcal{O}_{\mathcal{H}'}$ has the point-intersection property by Lemma 4.4. Therefore, by the induction hypothesis, x and y can be connected by a path in $P \cap \mathcal{H}'$, which implies that p and q can be connected in P.

Finally, observe that the intersection of a directed $\mathcal{O}$-halfspace with every $\mathcal{O}$-flat η is a directed $\mathcal{O}_\eta$-halfspace Lemma 5.4, and the lower-dimensional orientation set $\mathcal{O}_\eta$ has the point-intersection property by Lemma 4.4. Therefore, the intersection of P with every $\mathcal{O}$-flat η is path connected, which means that P is $\mathcal{O}$-path connected. □

Note that Theorem 5.5 cannot be extended to orientation sets without the point-intersection property. For example, if there is only one $\mathcal{O}$-line through o, then the union of several lines parallel to this $\mathcal{O}$-line is a disconnected directed $\mathcal{O}$-halfspace.

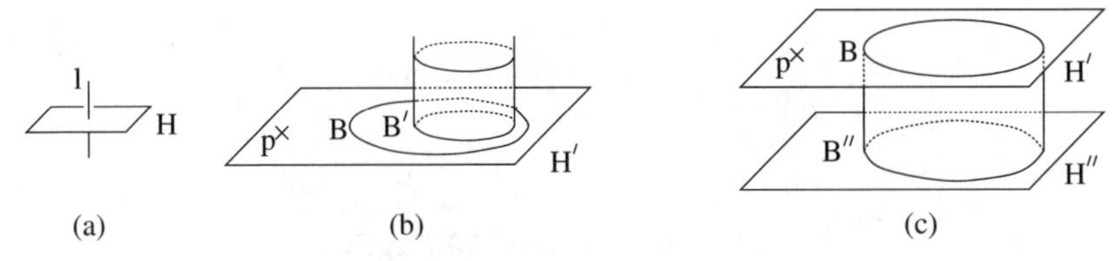

Fig. 5.8. Proof of Lemma 5.6

5.3 Boundary Convexity

We now characterize $\mathcal{O}$-halfspaces in terms of their boundaries. We begin by showing that all points in the boundary of an $\mathcal{O}$-halfspace are "infinitesimally close" to its interior; that is, an $\mathcal{O}$-halfspace is the closure of its interior.

Lemma 5.6. *Let P be an $\mathcal{O}$-halfspace and P_{int} be the interior of P. If the orientation set $\mathcal{O}$ has the point-intersection property, then the closure of P_{int} is P.*

Proof. We use induction on the dimension d to show that, for every point p in the boundary of P and every positive real number ϵ, there is an interior point within distance ϵ of p. By Lemma 4.2, we can select some $\mathcal{O}$-line l and an $\mathcal{O}$-hyperplane $\mathcal{H}$ that intersects l and does not contain it, as shown in Fig. 5.8a. Let $\mathcal{H}'$ be the $\mathcal{O}$-hyperplane through p parallel to $\mathcal{H}$, let Q be the intersection of P and $\mathcal{H}'$, and let Q_{int} be the interior of Q in the $(d-1)$-dimensional space $\mathcal{H}'$. If $d > 2$, then the set Q is a lower-dimensional $\mathcal{O}$-halfspace; hence, the closure of Q_{int} is Q by the induction hypothesis. If $d = 2$, then Q is empty, a ray or a line; therefore, we again conclude that the closure of Q_{int} is Q, which establishes the induction basis.

We select an interior point of Q, within distance $\epsilon/2$ of p, and a $(d-1)$-dimensional ball $B \subseteq Q$, centered at this point, such that the radius of B is at most $\epsilon/2$, as shown in Fig. 5.8b. We assume that the line l is vertical, and divide the rays parallel to l into up and down rays. For every point $q \in B$, the up or down ray with endpoint q is contained in P. Let U be the set of all points of B such that the up rays from them are in P, and let D be the set of all points of B such that the down rays from them are in P.

We consider two cases. First, suppose that the interior of U in the $(d-1)$-dimensional space $\mathcal{H}'$ is nonempty, which means that there is a $(d-1)$-dimensional ball $B' \subseteq U$, as shown in Fig. 5.8b. The union of the up rays from all points of B' forms a cylinder contained in P, and some interior points of this cylinder are within distance ϵ of p. Clearly, the interior of the cylinder is in the interior of P, which implies that some of P's interior points are within ϵ of p.

Second, suppose that the interior of U is empty. Then, every $(d-1)$-dimensional ball $B' \subseteq B$ contains a point of the set D, which implies that the closure of D is B. Let $\mathcal{H}''$ be a plane parallel to $\mathcal{H}'$ and located below $\mathcal{H}'$,

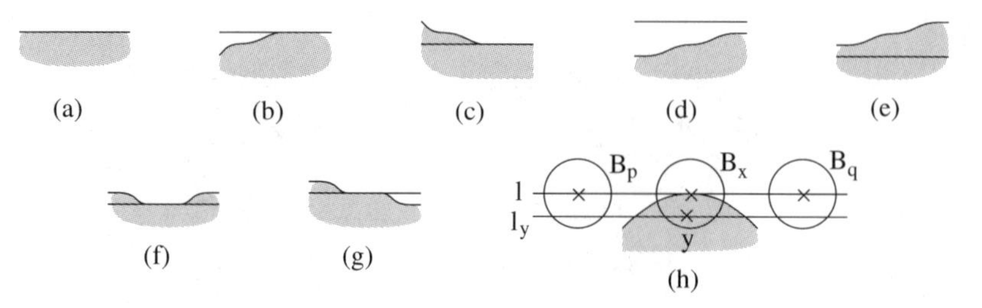

Fig. 5.9. Eight cases in the proof of Lemma 5.8

as shown in Fig. 5.8c. The intersection of $\mathcal{H}''$ with the down rays from all points of D forms a translate of D. This translate is in P and its closure is a $(d-1)$-dimensional ball, denoted B'', which is a translate of B. By the definition of $\mathcal{O}$-halfspaces, P is closed, which implies that B'' is in P. The union of the vertical segments joining B and B'' forms a cylinder, which is contained in P. The interior of this cylinder is in P's interior and some of the cylinder's interior points are within distance ϵ of p. □

The sets satisfying the closure property established in Lemma 5.6 are called **interior closed;** that is, a set is interior closed if it is identical to the closure of its interior.

We next give $\mathcal{O}$-convexity analogs of the following boundary-convexity characterization of standard halfspaces.

Proposition 5.7. *An interior-closed set is a halfspace if and only if its boundary is a nonempty convex set.*

We first generalize the "if" part of this characterization.

Lemma 5.8. *An interior-closed set is an $\mathcal{O}$-halfspace if its boundary is $\mathcal{O}$-convex.*

Proof. We consider an interior-closed set P with an $\mathcal{O}$-convex boundary, and show that its intersection with every $\mathcal{O}$-line l is empty, a ray or a line. The intersection of l with P's boundary may be empty, a point, a line segment, a ray or a line. If it is a line, then $P \cap l$ is also a line, as shown in Fig. 5.9a. If the intersection of l with P's boundary is a ray, then $P \cap l$ is a ray or a line, as shown in Fig. 5.9b,c. If l does not intersect P's boundary, then $P \cap l$ is empty or a line, as shown in Fig. 5.9d,e.

Finally, suppose that the intersection of l with P's boundary is a point or a line segment. We prove, by contradiction, that $P \cap l$ is a line or a ray, as shown in Fig. 5.9f,g. If not, then $P \cap l$ is a point or a line segment, as shown in Fig. 5.9h. We select points $p, q \in l$ that are outside of P, on different sides of the intersection, and consider equal-size balls, B_p and B_q, that are centered at these points and do not intersect P. We next choose a point x in

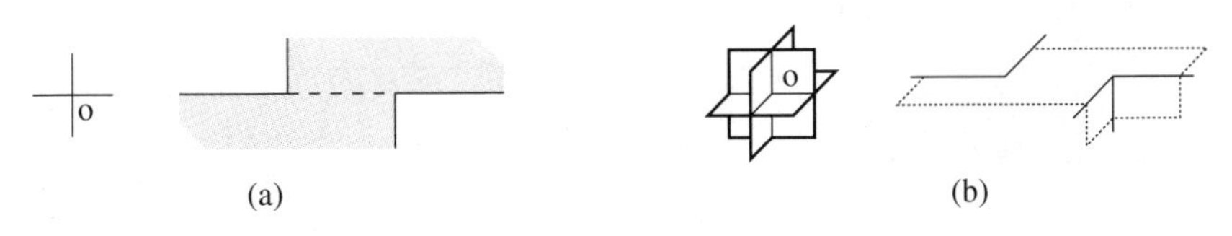

Fig. 5.10. $\mathcal{O}$-halfplane **(a)** and $\mathcal{O}$-halfspace **(b)** whose boundaries are not $\mathcal{O}$-convex

the intersection of l and P's boundary, and consider the ball B_x, centered at x, of the same size as B_p and B_q. Since P is interior closed, we can select a point $y \in B_x$ that is in the interior of P. Consider the line l_y through y parallel to l; this line intersects the balls B_p and B_q, which are outside of P. Since y is in the interior of P, the intersection of the $\mathcal{O}$-line l_y with the boundary of P is disconnected, contradicting the $\mathcal{O}$-convexity of P's boundary. □

The converse of Lemma 5.8 does not hold; that is, the boundary of an $\mathcal{O}$-halfspace may not be $\mathcal{O}$-convex. For example, the boundary of the $\mathcal{O}$-halfplane in Fig. 5.10a is not $\mathcal{O}$-convex, because its intersection with the dashed $\mathcal{O}$-line is disconnected; similarly, the boundary of the $\mathcal{O}$-halfspace in Fig. 5.10b is not $\mathcal{O}$-convex. We now characterize $\mathcal{O}$-halfspaces in terms of their boundaries.

Lemma 5.9 (Boundaries of $\mathcal{O}$-halfspaces). *An interior-closed set P is an $\mathcal{O}$-halfspace if and only if, for every $\mathcal{O}$-line l, one of the following two conditions holds:*

1. *The intersection of l with the boundary of P is connected.*
2. *The intersection of l with the boundary of P consists of two disconnected rays, and the segment of l between these rays is in P.*

Proof. Suppose that, for every $\mathcal{O}$-line l, one of the two conditions holds. We have shown in the proof of Lemma 5.8 that, if the intersection of l with the boundary of P satisfies Condition 1, then $P \cap l$ is empty, a ray or a line. Furthermore, if the intersection of l with P's boundary satisfies Condition 2, then $P \cap l$ is a line. We conclude that the intersection of P with every $\mathcal{O}$-line is empty, a ray or a line, which means that P is an $\mathcal{O}$-halfspace.

To prove the converse, suppose that P is an $\mathcal{O}$-halfspace and the intersection of an $\mathcal{O}$-line l with the boundary of P does not satisfy Condition 1, which means that it is disconnected. We show that, in this case, the intersection satisfies Condition 2.

Since the boundary is closed, we can select points $p, q \in l$ in P's boundary such that all points of l between p and q are not in the boundary. Since the intersection of l with P is connected, the segment of l between p and q is in P. We show, by contradiction, that all points of l outside of this segment are in P's boundary. Suppose that some point x is not in P's boundary; without loss of generality, we assume that x is to the left of p, and q is to the right of p, as shown in Fig. 5.11a.

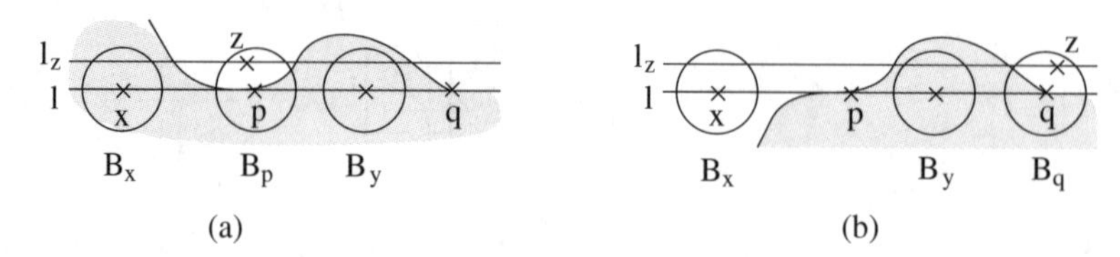

Fig. 5.11. Proof of Lemma 5.9

If x is an interior point of P, then there is a ball $B_x \subseteq P$ centered at x, as shown in Fig. 5.11a. Let $y \in l$ be some point between p and q, let $B_y \subseteq P$ be a ball centered at y, and let B_p be a ball, centered at p, such that B_p is no larger than B_x and B_y. Since p is in the boundary of P, there is a point $z \in B_p$ that is outside of P. Consider the $\mathcal{O}$-line l_z through z parallel to l, and note that this line intersects the balls B_x and B_y, which are in P. Since z is outside of P, the intersection of l_z with P is disconnected, contradicting the assumption that P is an $\mathcal{O}$-halfspace.

If x is an exterior point of P, we can select a ball B_x, centered at x, that does not intersect P, as shown in Fig. 5.11b. Let $y \in l$ be some point between p and q, let $B_y \subseteq P$ be a ball centered at y, and let B_q be a ball, centered at q, such that B_q is no larger than B_x and B_y. Since q is in the boundary of P, there is a point $z \in B_q$ that is outside of P. We again consider the $\mathcal{O}$-line l_z through z parallel to l. This line intersects the balls B_x and B_y, which implies that the intersection of l_z with P is nonempty, not a ray and not a line, contradicting the assumption that P is an $\mathcal{O}$-halfspace. □

Observe that, if the intersection of P's boundary with a line l consists of two rays, and the segment of l between these rays is in P, then l is wholly in P. This observation leads to a simpler version of Condition 2 in Lemma 5.9.

Theorem 5.10. *An interior-closed set P is an $\mathcal{O}$-halfspace if and only if, for every $\mathcal{O}$-line l, one of the following two conditions holds:*

1. *The intersection of l with the boundary of P is connected.*
2. *The $\mathcal{O}$-line l is wholly in P.*

The next result is a boundary-convexity property of directed $\mathcal{O}$-halfspaces.

Lemma 5.11. *The boundary of a directed $\mathcal{O}$-halfspace is $\mathcal{O}$-convex.*

Proof. Suppose that the boundary of a directed $\mathcal{O}$-halfspace P is not $\mathcal{O}$-convex, which means that its intersection with some $\mathcal{O}$-line l is disconnected. We can select points $p, q \in l$ that are in the boundary, and a point $x \in l$ between p and q that is not in the boundary, as shown in Fig. 5.12; for convenience, assume that p is to the left of x.

Since the intersection of P with l is connected, x is in the interior of P; hence, there is a ball $B_x \subseteq P$ centered at x. Either all left-directed rays or all

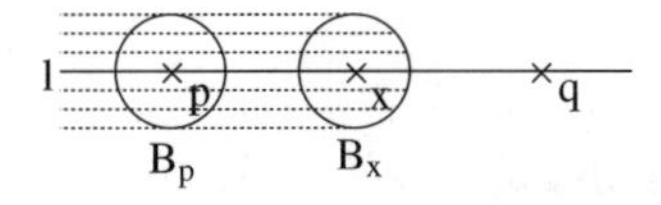

Fig. 5.12. Proof of Lemma 5.11

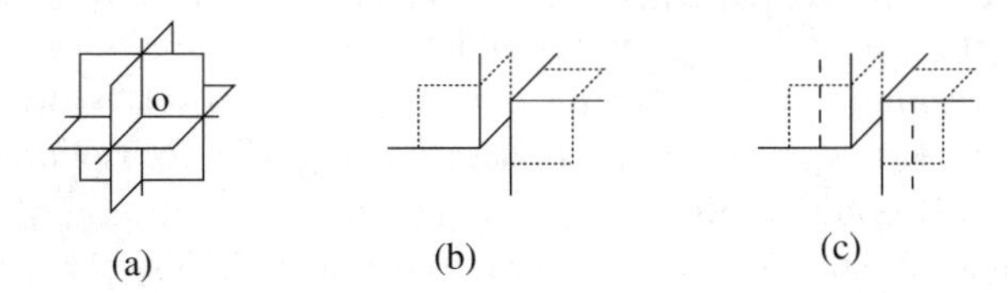

Fig. 5.13. Nondirected $\mathcal{O}$-halfspace with an $\mathcal{O}$-connected boundary

right-directed rays, with endpoints in B_x, are contained in P. Assume that the left-directed rays are in P; then, some ball B_p centered at p is wholly in P, which implies that p is in P's interior, thus yielding a contradiction. □

If an orientation set $\mathcal{O}$ has the point-intersection property, then the *boundary of any directed $\mathcal{O}$-halfspace is $\mathcal{O}$-connected,* which is a stronger result than Lemma 5.11; we delay its proof until Sect. 5.4. The converse does not hold, as shown in the following example.

Example: Nondirected $\mathcal{O}$-halfspace with an $\mathcal{O}$-connected boundary. Consider the $\mathcal{O}$-halfspace in Fig. 5.13b; it consists of two rectangular polyhedral angles, which touch each other along one of their facets. Its boundary is $\mathcal{O}$-connected for the orthogonal-orientation set in Fig. 5.13a; however, it is not directed, because the dashed rays in Fig. 5.13c, which are formed by its intersection with vertical $\mathcal{O}$-lines, have opposite directions.

5.4 Complementation

We give a condition under which the closure of the complement of an $\mathcal{O}$-halfspace is an $\mathcal{O}$-halfspace, and then show that the closure of the complement of a directed $\mathcal{O}$-halfspace is always a directed $\mathcal{O}$-halfspace.

We call the closure of the complement of a set the **closed complement.** Observe that the closed complement of an $\mathcal{O}$-halfspace may not be an $\mathcal{O}$-halfspace. For example, the closed complement of the $\mathcal{O}$-halfplane in Fig. 5.14b is not an $\mathcal{O}$-halfplane, since its intersection with the dashed $\mathcal{O}$-line in Fig. 5.14c is disconnected. As another example, the closed complement of the $\mathcal{O}$-halfspace in Fig. 5.10b is not an $\mathcal{O}$-halfspace.

Theorem 5.12 (Complements of $\mathcal{O}$-halfspaces). *The closed complement of an $\mathcal{O}$-halfspace P is an $\mathcal{O}$-halfspace if and only if the boundary of P is $\mathcal{O}$-convex.*

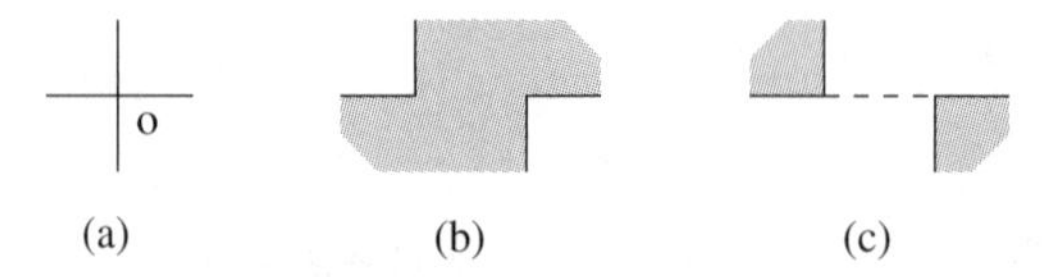

Fig. 5.14. $\mathcal{O}$-halfplane **(b)** whose closed complement **(c)** is not an $\mathcal{O}$-halfplane

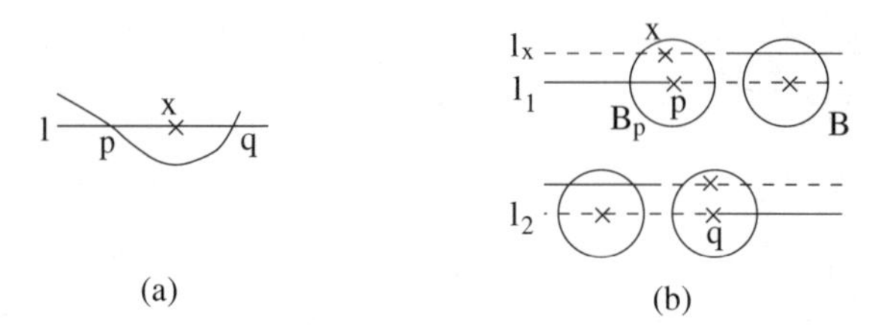

Fig. 5.15. Proofs of Theorem 5.12 **(a)** and Theorem 5.13 **(b)**

Proof. We denote the closed complement of P by Q. Note that Q is interior closed and its boundary is the same as the boundary of P. Thus, if P's boundary is $\mathcal{O}$-convex, then Q is an interior-closed set with an $\mathcal{O}$-convex boundary, which implies that Q is an $\mathcal{O}$-halfspace by Lemma 5.8.

We prove the converse by contradiction. Suppose that the boundary of P is not $\mathcal{O}$-convex and Q is an $\mathcal{O}$-halfspace. Then, there are points p, x and q on some $\mathcal{O}$-line l such that p and q are in the boundary, whereas x, located between p and q, is not in the boundary, as shown in Fig. 5.15a. Note that both p and q belong to P and to Q, whereas x is either in P's interior or in Q's interior. If x is in P, then the intersection of P with the $\mathcal{O}$-line l is disconnected, which means that P is not an $\mathcal{O}$-halfspace; similarly, if x is in Q, then Q is not an $\mathcal{O}$-halfspace. □

Theorem 5.13 (Complements of directed $\mathcal{O}$-halfspaces). *The closed complement of a directed $\mathcal{O}$-halfspace is a directed $\mathcal{O}$-halfspace.*

Proof. The boundary of a directed $\mathcal{O}$-halfspace P is $\mathcal{O}$-convex by Lemma 5.11; hence, the closed complement of P is an $\mathcal{O}$-halfspace by Theorem 5.12.

We show, by contradiction, that the closed complement of P is directed. Suppose that the intersection of the closed complement with two parallel $\mathcal{O}$-lines, l_1 and l_2, forms rays that have opposite directions. We denote the respective endpoints of these rays by p and q; in Fig. 5.15b, we show these rays by solid lines. Note that p and q are in P's boundary, and the dashed parts of l_1 and l_2 are in P's interior.

We select a point in the interior part of l_1 and a ball $B \subseteq P$ centered at this point, and let B_p be a ball of the same size as B centered at p. Since p is in the boundary of P, there is some point $x \in B_p$ that is not in P. Consider the $\mathcal{O}$-line l_x through x parallel to l_1, and note that the intersection of l_x

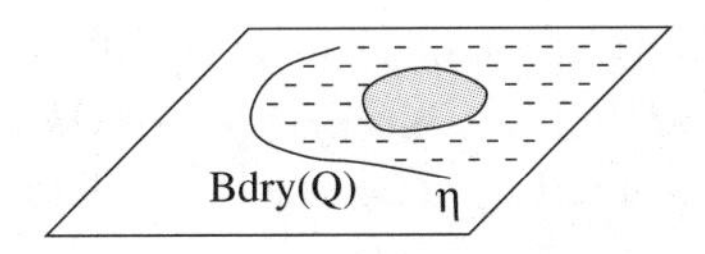

Fig. 5.16. Proof of Theorem 5.14

with P is empty, a ray or a line. Since x is not in P, and l_x intersects the ball $B \subseteq P$, we conclude that the intersection of l_x with P is a ray directed to the right. A similar construction with l_2 leads to an $\mathcal{O}$-line whose intersection with P is a ray directed to the left, as shown in Fig. 5.15b. This construction implies that the $\mathcal{O}$-halfspace P is not directed, yielding a contradiction. □

We apply Theorem 5.13 to show that the boundary of a directed $\mathcal{O}$-halfspace is $\mathcal{O}$-connected.

Theorem 5.14 (Boundaries of directed $\mathcal{O}$-halfspaces). *If an orientation set $\mathcal{O}$ has the point-intersection property, then the boundary of every directed $\mathcal{O}$-halfspace is $\mathcal{O}$-connected.*

Proof. To demonstrate that the boundary of a directed $\mathcal{O}$-halfspace P is $\mathcal{O}$-connected, we first show that the boundary is connected, and then use this result to prove that the intersection of the boundary with every $\mathcal{O}$-flat η is also connected.

We establish the connectedness of P's boundary by contradiction. Since the orientation set has the point-intersection property, P is connected by Theorem 5.5. Therefore, if the boundary of P is disconnected, the closed complement of P is also disconnected. On the other hand, the closed complement of P is a directed $\mathcal{O}$-halfspace by Theorem 5.13, contradicting the connectedness of directed $\mathcal{O}$-halfspaces.

We next show that the intersection of P's boundary, denoted by Bdry(P), with every $\mathcal{O}$-flat η is connected. Let Q be the intersection of P with η, and Bdry(Q) be the boundary of Q in the lower-dimensional space η; note that $\text{Bdry}(Q) \subseteq \text{Bdry}(P) \cap \eta$. The set Q is a directed $\mathcal{O}_\eta$-halfspace, and $\mathcal{O}_\eta$ has the point-intersection property by Lemma 4.4; hence, by the first part of the proof, Bdry(Q) is connected.

If $\text{Bdry}(P) \cap \eta$ is disconnected, then it has a component disconnected from the boundary of Q; we show this component by the shaded region in Fig. 5.16. Since this component is contained in Q, it is surrounded in the flat η by interior points of P; in Fig. 5.16, the interior points are shown by the dashed region.

We now consider the intersection of the closed complement of P with η. This intersection contains Bdry(Q) and the component of $\text{Bdry}(P) \cap \eta$ disconnected from Bdry(Q), but it does not contain any interior points of P; therefore, the intersection of the closed complement of P with η is disconnected. On the other hand, the closed complement of P is a directed $\mathcal{O}$-halfspace by Theorem 5.13, which implies that the intersection of the closed complement of P with the $\mathcal{O}$-flat η is connected by Theorem 5.5, yielding a contradiction. □

Summary

We have studied two generalizations of halfspaces, called $\mathcal{O}$-halfspaces and directed $\mathcal{O}$-halfspaces, and demonstrated that their properties are similar to those of standard halfspaces. In particular, we have characterized disconnected $\mathcal{O}$-halfspaces in terms of their connected components (Theorem 5.2), demonstrated that directed $\mathcal{O}$-halfspaces are $\mathcal{O}$-path connected and their boundaries are $\mathcal{O}$-connected (Theorems 5.5 and 5.14), presented a condition under which an interior-closed set is an $\mathcal{O}$-halfspace (Theorem 5.10), and described the closed complements of $\mathcal{O}$-halfspaces and directed $\mathcal{O}$-halfspaces (Theorems 5.12 and 5.13).

6

Strong Convexity

We extend strong $\mathcal{O}$-convexity to higher dimensions and discuss its basic properties (Sect. 6.1). Then, we characterize strongly $\mathcal{O}$-convex flats and derive a condition for the equivalence of two orientation sets (Sect. 6.2). Finally, we study strongly $\mathcal{O}$-convex halfspaces and characterize strongly $\mathcal{O}$-convex sets through halfspace intersections (Sect. 6.3).

6.1 Strongly $\mathcal{O}$-Convex Sets

We define $\mathcal{O}$-blocks and strongly $\mathcal{O}$-convex sets in a d-dimensional space $\mathcal{R}^d$, and then study their basic properties. Recall that planar strong $\mathcal{O}$-convexity has been defined through the $\mathcal{O}$-block visibility in Sect. 2.3.

The definition of $\mathcal{O}$-blocks in higher dimensions is based on the notion of an **$\mathcal{O}$-layer** of two points. Let $\mathcal{H}$ be an $\mathcal{O}$-hyperplane, let p and q be two points, let $\mathcal{H}_p$ be the $\mathcal{O}$-hyperplane through p parallel to $\mathcal{H}$, and let $\mathcal{H}_q$ be the $\mathcal{O}$-hyperplane through q parallel to $\mathcal{H}$. The closed layer of space between the $\mathcal{O}$-hyperplanes $\mathcal{H}_p$ and $\mathcal{H}_q$ is called the **$\mathcal{H}$-layer** of p and q.

We can define this $\mathcal{H}$-layer as the intersection of two halfspaces. Specifically, let P_p be the halfspace with boundary $\mathcal{H}_p$ that contains q, and P_q be the halfspace with boundary $\mathcal{H}_q$ that contains p; then, the $\mathcal{H}$-layer of p and q is $P_p \cap P_q$. As a special case, if $q \in \mathcal{H}_p$, then the hyperplane $\mathcal{H}_p$ is the $\mathcal{H}$-layer of p and q.

We can also describe the $\mathcal{H}$-layer by a linear inequality in Cartesian coordinates. Suppose that the equation for $\mathcal{H}_p$ is $a_1x_1 + a_2x_2 + \cdots + a_dx_d = b_p$ and the equation for $\mathcal{H}_q$ is $a_1x_1 + a_2x_2 + \cdots + a_dx_d = b_q$. Since $\mathcal{H}_p$ and $\mathcal{H}_q$ are parallel, their equations differ only by the values of their constant terms. If $b_p \leq b_q$, then the $\mathcal{H}$-layer of p and q is the set of points that satisfy the inequality $b_p \leq a_1x_1 + a_2x_2 + \cdots + a_dx_d \leq b_q$.

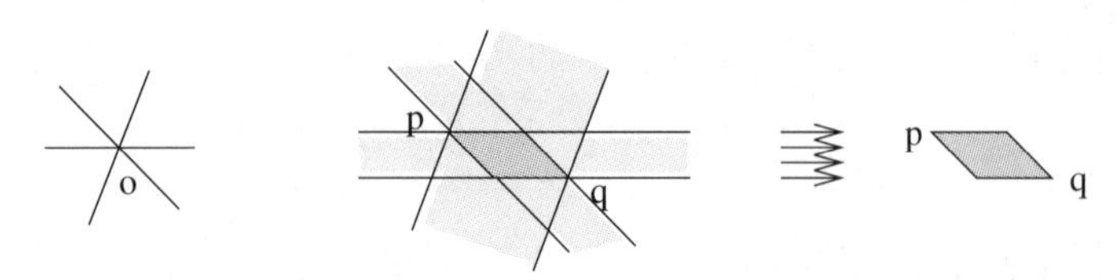

Fig. 6.1. Defining $\mathcal{O}$-block(p, q) through the intersection of $\mathcal{O}$-layers

Fig. 6.2. $\mathcal{O}$-blocks in three dimensions

Fig. 6.3. Strongly $\mathcal{O}$-convex sets

The **$\mathcal{O}$-block** of p and q is the intersection of all $\mathcal{O}$-layers of p and q:

$$\mathcal{O}\text{-block}(p, q) = \bigcap_{\mathcal{H} \in \mathcal{O}} \mathcal{H}\text{-layer}(p, q).$$

In other words, a point is in the $\mathcal{O}$-block of p and q if, for every $\mathcal{O}$-hyperplane $\mathcal{H}$, this point is between $\mathcal{H}_p$ and $\mathcal{H}_q$. Note that this definition of $\mathcal{O}$-blocks is equivalent to their definition in the proof of Theorem 4.16.

In two dimensions, we can describe $\mathcal{O}$-blocks in the same way; a planar layer is the set of points between two parallel lines, and the $\mathcal{O}$-block of p and q is the intersection of all $\mathcal{O}$-layers of p and q. This view of $\mathcal{O}$-blocks is equivalent to their definition in Sect. 2.3, as illustrated in Fig. 6.1.

We give examples of three-dimensional $\mathcal{O}$-blocks in Fig. 6.2. For the orientation set in Fig. 6.2a, $\mathcal{O}$-blocks are parallelepipeds. The orientation set in Fig. 6.2b gives rise to more complex $\mathcal{O}$-blocks.

The definition of strong $\mathcal{O}$-convexity in higher dimensions is identical to its definition in two dimensions; that is, *a set is strongly $\mathcal{O}$-convex if, for every two of its points, their $\mathcal{O}$-block is wholly in the set.* In Fig. 6.3, we show strongly $\mathcal{O}$-convex polytopes, whose facets are parallel to elements of $\mathcal{O}$.

In Lemma 2.9, we have given six basic properties of planar strong $\mathcal{O}$-convexity. We can readily generalize Properties 1–3 to higher dimensions; specifically, every translate of a strongly $\mathcal{O}$-convex set is strongly $\mathcal{O}$-convex, the intersection of strongly $\mathcal{O}$-convex sets is strongly $\mathcal{O}$-convex, and every strongly $\mathcal{O}$-convex set is convex. Property 4 also holds in higher dimensions, as shown in Corollary 6.3.

Property 5 holds in one direction; that is, if $\mathcal{O}_1$ and $\mathcal{O}_2$ have identical closures, then strong $\mathcal{O}_1$-convexity is equivalent to strong $\mathcal{O}_2$-convexity, as shown in Corollary 6.3. On the other hand, strong $\mathcal{O}_1$-convexity may be equivalent to strong $\mathcal{O}_2$-convexity even if the closures of $\mathcal{O}_1$ and $\mathcal{O}_2$ are distinct, as shown in Sect. 6.2.

The analog of Property 6 holds in higher dimensions for *finite orientation sets,* as shown in Corollary 6.17; that is, for finite $\mathcal{O}$, a polytope is strongly $\mathcal{O}$-convex if and only if it is convex and its facets are parallel to elements of $\mathcal{O}$. For an infinite orientation set, a polytope may be strongly $\mathcal{O}$-convex even if its facets are not parallel to elements of $\mathcal{O}$, as shown in Sect. 6.3.

Lemma 6.1. *Every convex set is strongly $\mathcal{O}$-convex if and only if every line is strongly $\mathcal{O}$-convex.*

Proof. Clearly, if every convex set is strongly $\mathcal{O}$-convex, then every line is strongly $\mathcal{O}$-convex. To prove the converse, suppose that every line is strongly $\mathcal{O}$-convex. Then, for every two points p and q, their $\mathcal{O}$-block is the line segment joining them. If the $\mathcal{O}$-block were a superset of this segment, the line through p and q would not be strongly $\mathcal{O}$-convex. Therefore, in this case, strong $\mathcal{O}$-convexity is equivalent to standard convexity. $\square$

Next, we give several basic properties of $\mathcal{O}$-blocks.

Lemma 6.2. *For every two points p and q:*

1. *If $\mathcal{O}_1 \subseteq \mathcal{O}_2$, then $\mathcal{O}_2$-block$(p,q) \subseteq \mathcal{O}_1$-block$(p,q)$.*
2. *If $\mathcal{O}_2$ is the closure of $\mathcal{O}_1$, then $\mathcal{O}_1$-block$(p,q) = \mathcal{O}_2$-block(p,q).*
3. *The $\mathcal{O}$-block of p and q is strongly $\mathcal{O}$-convex.*
4. *The $\mathcal{O}$-block of p and q is the union of all simple $\mathcal{O}$-connected segments joining p and q.*

Proof.

(1) If $\mathcal{O}_1 \subseteq \mathcal{O}_2$, then every $\mathcal{O}_1$-layer is an $\mathcal{O}_2$-layer, which implies that $\mathcal{O}_2$-block(p,q) is a subset of $\mathcal{O}_1$-block(p,q).

(2) If $\mathcal{O}_2$ is the closure of $\mathcal{O}_1$, then $\mathcal{O}_1 \subseteq \mathcal{O}_2$, which implies that $\mathcal{O}_2$-block$(p,q) \subseteq \mathcal{O}_1$-block$(p,q)$. We prove the converse inclusion by showing that, for every $\mathcal{O}_2$-layer of p and q, $\mathcal{O}_1$-block(p,q) is a subset of this layer; that is, if a point x is outside of the layer, then it is also outside of $\mathcal{O}_1$-block(p,q).

Let $\mathcal{H}$-layer(p,q) be an $\mathcal{O}_2$-layer, with boundary hyperplanes $\mathcal{H}_p$ (through p) and $\mathcal{H}_q$ (through q), and let x be a point outside of $\mathcal{H}$-layer(p,q). Without loss of generality, assume that either $\mathcal{H}_p = \mathcal{H}_q$ or $\mathcal{H}_q$ is between x and $\mathcal{H}_p$, as shown in Fig. 6.4a.

If $\mathcal{H}_q$ is an $\mathcal{O}_1$-hyperplane, then $\mathcal{O}_1$-block$(p,q) \subseteq \mathcal{H}$-layer$(p,q)$, which implies that x is outside of $\mathcal{O}_1$-block(p,q). If $\mathcal{H}_q$ is not an $\mathcal{O}_1$-hyperplane, then there is a sequence of $\mathcal{O}_1$-hyperplanes through q convergent to $\mathcal{H}_q$. For

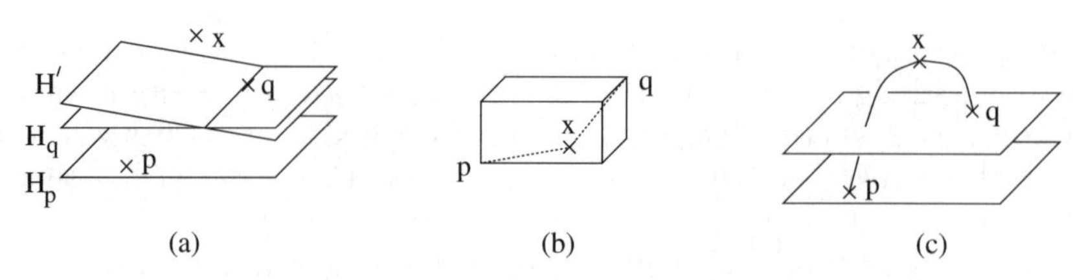

Fig. 6.4. Proof of Lemma 6.2

some hyperplane $\mathcal{H}'$ of this sequence, p and x are "on different sides" of $\mathcal{H}'$, as shown in Fig. 6.4a. The $\mathcal{H}'$-layer of p and q is an $\mathcal{O}_1$-layer, and x is outside of this layer; therefore, we again conclude that x is outside of $\mathcal{O}_1$-block(p, q).

(3) We have to show that, for every two points x and y in $\mathcal{O}$-block(p, q), their $\mathcal{O}$-block is wholly in $\mathcal{O}$-block(p, q). Note that, for every $\mathcal{O}$-hyperplane $\mathcal{H}$, the points x and y are in the $\mathcal{H}$-layer of p and q, which implies that $\mathcal{H}$-layer$(x, y) \subseteq \mathcal{H}$-layer$(p, q)$. Since the $\mathcal{O}$-block of two points is the intersection of all their $\mathcal{O}$-layers, we conclude that $\mathcal{O}$-block$(x, y) \subseteq \mathcal{O}$-block$(p, q)$.

(4) First, we show that, if a point x is in $\mathcal{O}$-block(p, q), then there is an $\mathcal{O}$-connected segment through x joining p and q. Consider the polygonal segment $c[p, q]$ formed by two *line segments,* $c[p, x]$ and $c[x, q]$, as shown in Fig. 6.4b. Observe that, if some $\mathcal{O}$-hyperplane intersects both $c[p, x]$ and $c[x, q]$, then it contains x. By Lemma 4.14, this observation implies that $c[p, q]$ is an $\mathcal{O}$-connected segment.

Next, we consider a point x outside of $\mathcal{O}$-block(p, q) and show that any segment $c[p, q]$ through x is not $\mathcal{O}$-connected. Since x is not in the $\mathcal{O}$-block of p and q, it is outside of some $\mathcal{O}$-layer of p and q, as shown in Fig. 6.4c. If $c[p, q]$ is a simple curvilinear segment containing x, then its intersection with a boundary hyperplane of this $\mathcal{O}$-layer is disconnected, which means that it is not $\mathcal{O}$-connected. □

Parts 1 and 2 of Lemma 6.2 immediately imply the following results, which show that *we may restrict attention to the study of strong $\mathcal{O}$-convexity for closed orientation sets.*

Corollary 6.3.

1. *If $\mathcal{O}_1 \subseteq \mathcal{O}_2$, then every strongly $\mathcal{O}_1$-convex set is strongly $\mathcal{O}_2$-convex.*
2. *If $\mathcal{O}_2$ is the closure of $\mathcal{O}_1$, then strong $\mathcal{O}_1$-convexity is equivalent to strong $\mathcal{O}_2$-convexity; that is, a set is strongly $\mathcal{O}_1$-convex if and only if it is strongly $\mathcal{O}_2$-convex.*

Part 3 of Lemma 6.2 implies that $\mathcal{O}$-block(p, q) *is the minimal strongly $\mathcal{O}$-convex set containing p and q.* In other words, it is the intersection of all strongly $\mathcal{O}$-convex sets that contain p and q, which means that it is the strong $\mathcal{O}$-hull of p and q. Finally, Part 4 enables us to characterize strongly $\mathcal{O}$-convex sets in terms of the $\mathcal{O}$-connected curves defined in Sect. 4.3.

Theorem 6.4.

1. *A set is strongly $\mathcal{O}$-convex if and only if every simple $\mathcal{O}$-connected segment joining two of its points is wholly in the set.*
2. *A set is strongly $\mathcal{O}$-convex if and only if its intersection with every simple $\mathcal{O}$-connected curve is connected.*

Proof. The first statement readily follows from the observation that the $\mathcal{O}$-block of two points is the union of the $\mathcal{O}$-connected segments joining these points, as shown in Lemma 6.2.

To prove the second statement, we first consider the intersection of a strongly $\mathcal{O}$-convex set P with an $\mathcal{O}$-connected curve c, and show that this intersection is path connected. Let p and q be two arbitrary points in $P \cap c$; since the segment of c between p and q is contained in P, we conclude that there is a path joining p and q in $P \cap c$.

Next, we consider a set P that is not strongly $\mathcal{O}$-convex, and show that its intersection with some $\mathcal{O}$-connected curve is disconnected. By the first statement of the theorem, there is some simple $\mathcal{O}$-connected segment $c[p, q]$, joining two points of P, that is not wholly in P. By Lemma 4.12, there is a simple $\mathcal{O}$-connected curve c such that $c[p, q]$ is a segment of c. The intersection of P with this curve is disconnected. □

6.2 Strongly $\mathcal{O}$-Convex Flats

We explore the properties of strongly $\mathcal{O}$-convex flats, derive a necessary and sufficient condition for the equivalence of strong $\mathcal{O}$-convexity with respect to different orientation sets, and establish the strong $\mathcal{O}$-convexity of the affine hull of a strongly $\mathcal{O}$-convex set.

Lemma 6.5. *Every $\mathcal{O}$-flat is strongly $\mathcal{O}$-convex.*

Proof. If points p and q are in an $\mathcal{O}$-flat, then the $\mathcal{O}$-block of p and q is contained in this $\mathcal{O}$-flat, because $\mathcal{O}$-block(p, q) is a subset of every $\mathcal{O}$-hyperplane through p and q, and the $\mathcal{O}$-flat is the intersection of these $\mathcal{O}$-hyperplanes. □

If $\mathcal{O}$ is a closed countable set, the converse of Lemma 6.5 also holds, as shown in Theorem 6.7; that is, every strongly $\mathcal{O}$-convex flat is an $\mathcal{O}$-flat. If $\mathcal{O}$ is not closed, all hyperplanes in the closure of $\mathcal{O}$ are strongly $\mathcal{O}$-convex, although some of them are not $\mathcal{O}$-hyperplanes. For closed uncountable $\mathcal{O}$, flats may also be strongly $\mathcal{O}$-convex even if they are not $\mathcal{O}$-flats, as shown in the following example.

Example: Strongly $\mathcal{O}$-convex flats that are not $\mathcal{O}$-flats.
Let $\mathcal{O}_{\text{sc}}$ be the orientation set in three dimensions that includes all planes

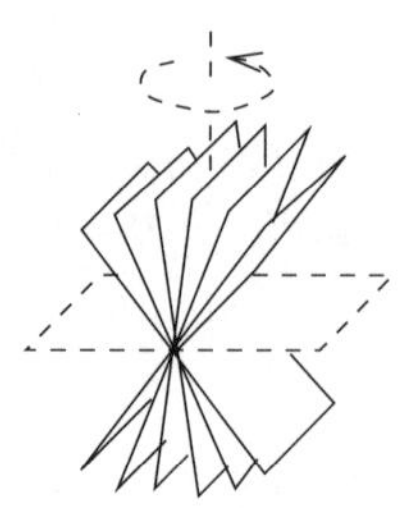

Fig. 6.5. Construction of the orientation set $\mathcal{O}_{sc}$

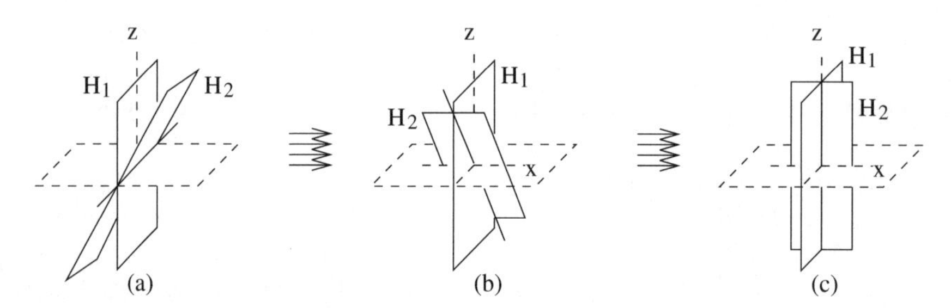

Fig. 6.6. Demonstrating that all lines are $\mathcal{O}_{sc}$-lines

through o whose angles with the "horizontal" plane are at least $\pi/3$, where any plane through o can serve as the horizontal. We illustrate the construction of $\mathcal{O}_{sc}$ in Fig. 6.5, where the horizontal plane is shown by dashed lines. The set comprises the (uncountably many) planes shown by solid lines and all rotations of these planes around the vertical axis. The subscript "sc" stands for "standard convexity" since we show that strong $\mathcal{O}_{sc}$-convexity is equivalent to standard convexity.

We show that every line through o is the intersection of two planes of $\mathcal{O}_{sc}$; an informal proof is illustrated in Fig. 6.6, where H_1 and H_2 are elements of $\mathcal{O}_{sc}$. In Fig. 6.6a, the intersection of H_1 and H_2 is a horizontal line. Now suppose that we rotate H_2 around the vertical axis z, until it reaches the position shown in Fig. 6.6b, and then rotate H_2 around the horizontal axis x, until it becomes, as shown in Fig. 6.6c. At all times, H_2 is a plane in $\mathcal{O}$, and the intersection of H_1 and H_2 is always a line, whose position continuously changes from horizontal to vertical. Since every rotation around the vertical axis z maps $\mathcal{O}_{sc}$ into itself, we conclude that every line through o is the intersection of two elements of $\mathcal{O}_{sc}$.

Since translates of elements of $\mathcal{O}_{sc}$ are $\mathcal{O}_{sc}$-planes, every line is the intersection of two $\mathcal{O}_{sc}$-planes; therefore, every line is strongly $\mathcal{O}_{sc}$-convex, which implies that strong $\mathcal{O}_{sc}$-convexity is equivalent to standard convexity by Lemma 6.1. Thus, all planes are strongly $\mathcal{O}_{sc}$-convex, although some of them are not $\mathcal{O}_{sc}$-planes.

Note that, if we define $\mathcal{O}'_{\mathrm{sc}}$ as the set of planes through o whose angles with some *vertical plane* are at least $\pi/3$, then strong $\mathcal{O}'_{\mathrm{sc}}$-convexity is also equivalent to standard convexity. This example shows that two distinct closed orientation sets may give rise to the same strong convexity, which means that Property 5 of Lemma 2.9 does not hold in three dimensions.

We next characterize strongly $\mathcal{O}$-convex flats in terms of $\mathcal{O}$-flats.

Theorem 6.6. *For a closed orientation set $\mathcal{O}$, a flat η is strongly $\mathcal{O}$-convex if and only if, for every two points of η, there is an $\mathcal{O}$-flat through them that is contained in η.*

Proof. Suppose that, for every two points $p, q \in \eta$, there is an $\mathcal{O}$-flat $H \subseteq \eta$ through p and q. Since H is strongly $\mathcal{O}$-convex, we conclude that $\mathcal{O}$-block$(p,q) \subseteq H \subseteq \eta$. Thus, for every two points of η, their $\mathcal{O}$-block is in η, which means that η is strongly $\mathcal{O}$-convex.

Suppose, conversely, that η is strongly $\mathcal{O}$-convex and consider two of its points, p and q. Let H be the $\mathcal{O}$-flat formed by the intersection of all $\mathcal{O}$-hyperplanes through p and q. We show, by contradiction, that $H \subseteq \eta$.

If H is not in η, then $H \cap \eta$ is a strongly $\mathcal{O}$-convex flat whose dimension is less than the dimension of H. Let x be the midpoint of the line segment joining p and q. Since $\mathcal{O}$-block$(p,q) \subseteq H \cap \eta$ and the dimension of $H \cap \eta$ is less than the dimension of H, we conclude that, for every ball B_x centered at x, $H \cap B_x \not\subseteq \mathcal{O}$-block$(p,q)$, which implies that there is an $\mathcal{O}$-layer of p and q that does not contain $H \cap B_x$.

If a layer does not contain $H \cap B_x$, then both boundary hyperplanes of this layer intersect B_x and do not contain H. Therefore, we can select a sequence of $\mathcal{O}$-hyperplanes through p that do not contain H, such that the distances from these hyperplanes to x converge to zero. We choose a convergent subsequence of this sequence and take its limit, thus getting an $\mathcal{O}$-hyperplane through p and q that does not contain H, which contradicts the definition of H. □

We have noted that $\mathcal{O}$-flats are strongly $\mathcal{O}$-convex. We now show that, for finite and closed countably infinite orientation sets, the converse also holds.

Theorem 6.7. *Suppose that $\mathcal{O}$ is a closed countable set; then, a flat is strongly $\mathcal{O}$-convex if and only if it is an $\mathcal{O}$-flat.*

Proof. We consider a flat η that is not an $\mathcal{O}$-flat and show that η is not strongly $\mathcal{O}$-convex. We denote the dimension of η by k; note that the dimension of an $\mathcal{O}$-flat contained in η is at most $k-1$.

Let p be some point of η; the set of $\mathcal{O}$-hyperplanes through p is countable, and the intersections of these hyperplanes form countably many $\mathcal{O}$-flats. Thus, there are only countably many $\mathcal{O}$-flats through p contained in η. Since the dimension of these flats is at most $k-1$, they do not cover η. Thus, there is a point $q \in \eta$ such that no $\mathcal{O}$-flat through p and q is wholly in η, which implies that η is not strongly $\mathcal{O}$-convex by Theorem 6.6. □

For points and lines, the analog of Theorem 6.7 holds even when an orientation set is uncountable.

Theorem 6.8. *Suppose that $\mathcal{O}$ is a closed orientation set; then, a point or line is strongly $\mathcal{O}$-convex if and only if it is an $\mathcal{O}$-flat.*

Proof. Every $\mathcal{O}$-flat is strongly $\mathcal{O}$-convex by Lemma 6.5, which establishes the "if" direction. We prove the "only if" direction for a point and then for a line.

Suppose that a point p is strongly $\mathcal{O}$-convex and consider $\mathcal{O}$-block(p, p). By definition, this $\mathcal{O}$-block is the intersection of all $\mathcal{O}$-hyperplanes through p. Since p is strongly $\mathcal{O}$-convex, $\mathcal{O}$-block(p, p) is contained in p. Therefore, the point p is the intersection of $\mathcal{O}$-hyperplanes, which means that it is an $\mathcal{O}$-flat.

Now suppose that a line l is strongly $\mathcal{O}$-convex, and let p and q be distinct points of l. By Theorem 6.6, there is an $\mathcal{O}$-flat through p and q contained in l. Since the only flat through p and q that is contained in l is l itself, we conclude that the line l is an $\mathcal{O}$-flat. □

If we combine Lemma 6.1 and Theorem 6.8, we get the following result.

Corollary 6.9. *Suppose that $\mathcal{O}$ is a closed orientation set; then, every convex set is strongly $\mathcal{O}$-convex if and only if every line is an $\mathcal{O}$-line.*

Note that we cannot extend Corollary 6.9 to nonclosed orientation sets. For example, consider three-dimensional space, let l be a line through o, and let $\mathcal{O}$ include all planes through o that do not contain l. Since the closure of $\mathcal{O}$ includes all planes, strong $\mathcal{O}$-convexity is equivalent to standard convexity, which implies that l is strongly $\mathcal{O}$-convex. On the other hand, l is not an $\mathcal{O}$-line, since it is not the intersection of any $\mathcal{O}$-planes.

For a given orientation set $\mathcal{O}$, we define $\widetilde{\mathcal{O}}$ as the set of all strongly $\mathcal{O}$-convex hyperplanes through o. For example, consider the three-dimensional orientation set $\mathcal{O}_{\text{sc}}$ illustrated in Fig. 6.5. We have shown that all planes are strongly $\mathcal{O}_{\text{sc}}$-convex; thus, $\widetilde{\mathcal{O}}_{\text{sc}}$ contains all planes through o. We discuss the notion of strong $\widetilde{\mathcal{O}}$-convexity, which is strong convexity with respect to the orientation set $\widetilde{\mathcal{O}}$. Observe that $\mathcal{O} \subseteq \widetilde{\mathcal{O}}$, which implies that every strongly $\mathcal{O}$-convex set is strongly $\widetilde{\mathcal{O}}$-convex by Corollary 6.3. We show that the converse also holds; that is, every strongly $\widetilde{\mathcal{O}}$-convex set is strongly $\mathcal{O}$-convex.

Theorem 6.10. *Suppose that $\mathcal{O}_1$ and $\mathcal{O}_2$ are orientation sets through the same point o.*

1. *Strong $\mathcal{O}_1$-convexity is equivalent to strong $\widetilde{\mathcal{O}}_1$-convexity.*
2. *If strong $\mathcal{O}_1$-convexity is equivalent to strong $\mathcal{O}_2$-convexity, then $\mathcal{O}_1 \subseteq \widetilde{\mathcal{O}}_2$.*

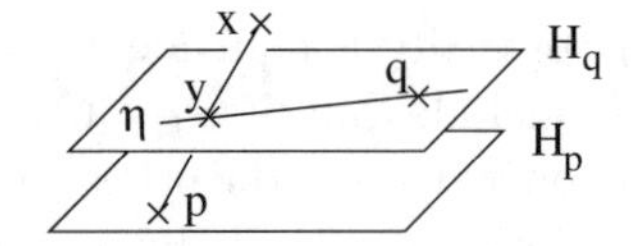

Fig. 6.7. Proof of Theorem 6.10

3. Strong $\mathcal{O}_1$-convexity is equivalent to strong $\mathcal{O}_2$-convexity if and only if $\widetilde{\mathcal{O}}_1 = \widetilde{\mathcal{O}}_2$.

Proof.

(1) We prove the equivalence by showing that, for every two points p and q, $\mathcal{O}_1$-block$(p, q) = \widetilde{\mathcal{O}}_1$-block$(p, q)$. By Lemma 6.2, we can assume that $\mathcal{O}_1$ is closed without loss of generality.

Since $\mathcal{O}_1 \subseteq \widetilde{\mathcal{O}}_1$, we conclude that $\widetilde{\mathcal{O}}_1$-block$(p, q) \subseteq \mathcal{O}_1$-block$(p, q)$ by Lemma 6.2. To prove the converse inclusion, we show that, for every $\widetilde{\mathcal{O}}_1$-layer of p and q, $\mathcal{O}_1$-block(p, q) is a subset of this layer; that is, if a point x is outside of the $\widetilde{\mathcal{O}}_1$-layer, then it is also outside of $\mathcal{O}_1$-block(p, q).

Consider an $\widetilde{\mathcal{O}}_1$-layer of p and q, with boundary hyperplanes $\mathcal{H}_p$ (through p) and $\mathcal{H}_q$ (through q), and let x be a point outside of this layer, as shown in Fig. 6.7. Since $\mathcal{H}_p$ and $\mathcal{H}_q$ are $\widetilde{\mathcal{O}}_1$-hyperplanes, they are strongly $\mathcal{O}_1$-convex.

First, suppose that $\mathcal{H}_p = \mathcal{H}_q$; that is, q is in $\mathcal{H}_p$. Then, the $\mathcal{O}_1$-block of p and q is a subset of $\mathcal{H}_p$, because $\mathcal{H}_p$ is strongly $\mathcal{O}_1$-convex; therefore, x is outside of $\mathcal{O}_1$-block(p, q).

Next, suppose that $\mathcal{H}_p$ and $\mathcal{H}_q$ are distinct hyperplanes. Without loss of generality, assume that $\mathcal{H}_q$ is between x and $\mathcal{H}_p$, as shown in Fig. 6.7. Then, the segment joining p and x intersects $\mathcal{H}_q$; we denote the intersection point by y. By Theorem 6.6, since $\mathcal{O}_1$ is closed, there is an $\mathcal{O}_1$-flat η through q and y that is contained in $\mathcal{H}_q$.

Observe that $x \notin \mathcal{H}_q$, which implies that $x \notin \eta$. Also note that, by the definition of $\mathcal{O}_1$-flats, η is the intersection of several $\mathcal{O}_1$-hyperplanes. If x belonged to all these hyperplanes, then x would be in η, leading to a contradiction. We conclude that there is an $\mathcal{O}_1$-hyperplane $\mathcal{H}_1$ that contains η and does not contain x.

We next show that $\mathcal{H}_1$ does not contain p. Note that y is in $\mathcal{H}_1$ and that the points p, y and x are on a line. If $\mathcal{H}_1$ contained p, it would contain the line through p and y, which implies that it would contain x, leading to a contradiction. We conclude that x and p are "on different sides" of $\mathcal{H}_1$, which means that x is outside of the $\mathcal{H}_1$-layer of p and q. Since the $\mathcal{H}_1$-layer is an $\mathcal{O}_1$-layer, we conclude that x is outside of $\mathcal{O}_1$-block(p, q).

(2) If strong $\mathcal{O}_1$-convexity is equivalent to strong $\mathcal{O}_2$-convexity, then $\widetilde{\mathcal{O}}_2$ contains all strongly $\mathcal{O}_1$-convex hyperplanes through o. Since every $\mathcal{O}_1$-hyperplane is strongly $\mathcal{O}_1$-convex, we conclude that $\mathcal{O}_1 \subseteq \widetilde{\mathcal{O}}_2$.

(3) Since strong $\mathcal{O}_1$-convexity is equivalent to strong $\widetilde{\mathcal{O}}_1$-convexity and the same holds for $\mathcal{O}_2$, we conclude that, if $\widetilde{\mathcal{O}}_1 = \widetilde{\mathcal{O}}_2$, then strong $\mathcal{O}_1$-convexity is equivalent to strong $\mathcal{O}_2$-convexity. On the other hand, if strong $\mathcal{O}_1$-convexity is equivalent to strong $\mathcal{O}_2$-convexity, then a hyperplane is strongly $\mathcal{O}_1$-convex if and only if it is strongly $\mathcal{O}_2$-convex; therefore, $\widetilde{\mathcal{O}}_1 = \widetilde{\mathcal{O}}_2$ by definition. □

We conclude that $\widetilde{\mathcal{O}}$ is the maximal orientation set for which strong convexity is equivalent to strong $\mathcal{O}$-convexity. Thus, for every orientation set $\mathcal{O}$, there is a unique maximal orientation set that gives rise to the same strong convexity. On the other hand, there may not be a unique minimal set for which strong convexity is equivalent to strong $\mathcal{O}$-convexity. For example, suppose that $\mathcal{O}$ is an orientation set in two dimensions that contains all lines through o. By Lemma 2.9, strong convexity with respect to some orientation set $\mathcal{O}_1$ is equivalent to strong $\mathcal{O}$-convexity if and only if the closure of $\mathcal{O}_1$ is $\mathcal{O}$, and the collection of all such orientation sets does not have a unique minimal element. In fact, it does not have any minimal elements; for every set $\mathcal{O}_1$ whose closure is $\mathcal{O}$, we can construct a proper subset of $\mathcal{O}_1$ whose closure is also $\mathcal{O}$ by removing any line from $\mathcal{O}_1$.

Since strong convexity for any orientation set is equivalent to strong convexity for its closure, Theorems 6.7 and 6.10 yield the following result.

Corollary 6.11.

1. *If $\mathcal{O}$ is a closed countable orientation set, then $\widetilde{\mathcal{O}} = \mathcal{O}$.*
2. *For every $\mathcal{O}$, the set $\widetilde{\mathcal{O}}$ is closed.*

We now show that the affine hull of a strongly $\mathcal{O}$-convex set is strongly $\mathcal{O}$-convex. The **affine hull** of a set P is the minimal flat that contains P; in other words, it is the intersection of all flats that contain P. For example, the affine hull of a line segment is a line, the affine hull of a triangle is a two-dimensional plane, and the affine hull of a ball is the whole space.

Since the affine hull of P is a lower-dimensional space, we can speak of the interior of P within this space, called the **relative interior.** Recall that a point is in the interior of P if there is a ball, centered at this point, that is wholly in P. If the affine hull of P is a k-dimensional flat, then a point is in the relative interior of P if there is a k-dimensional ball, centered at this point, that is wholly in P. For instance, suppose that P is a triangle in a three-dimensional space, and H is the plane that contains it. The interior of P in three dimensions is empty, whereas the relative interior of P in H includes all its points except its sides. We use the following property of relative interiors; for example, consult the text of Grünbaum, Klee, Perles, and Shephard [15].

Proposition 6.12. *If P is a convex set and η is its affine hull, then the relative interior of P in η is nonempty.*

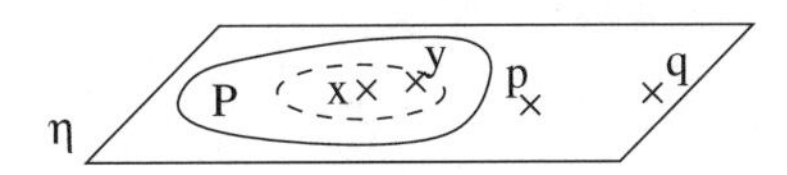

Fig. 6.8. Proof of Lemma 6.13

The next result is a key property of the affine hulls of strongly $\mathcal{O}$-convex sets, which will enable us to characterize strongly $\mathcal{O}$-convex sets in terms of halfspace intersections.

Lemma 6.13. *The affine hull of a strongly $\mathcal{O}$-convex set is strongly $\mathcal{O}$-convex.*

Proof. Let P be a strongly $\mathcal{O}$-convex set and η be its affine hull, as shown in Fig. 6.8. Since P is convex, the relative interior of P in η is nonempty; hence, we can choose an interior point x in P and a ball $B_x \subseteq P$ centered at x. Note that B_x is a ball in the space η rather than in the entire d-dimensional space; it is shown by dashes in Fig. 6.8.

We have to show that, for every two points p and q of η, their $\mathcal{O}$-block is in η. Let y be a point in B_x such that the line through x and y is parallel to the line through p and q. The $\mathcal{O}$-block of x and y is in P; therefore, it is in η. We next observe that $\mathcal{O}$-block(p, q) and $\mathcal{O}$-block(x, y) are *homothetic*, which means that we can transform one of them into the other by an increase or decrease of all distances in the same ratio. Since $\mathcal{O}$-block(x, y) is in η, this homothetic transformation implies that $\mathcal{O}$-block(p, q) is also in η. □

6.3 Strongly $\mathcal{O}$-Convex Halfspaces

We consider strongly $\mathcal{O}$-convex halfspaces and show that their role in strong $\mathcal{O}$-convexity is similar to the role of halfspaces in standard convexity. We begin by characterizing strongly $\mathcal{O}$-convex halfspaces through their boundaries.

Theorem 6.14. *A halfspace is strongly $\mathcal{O}$-convex if and only if its boundary is a strongly $\mathcal{O}$-convex hyperplane.*

Proof. Let P be a halfspace and let its boundary hyperplane $\mathcal{H}$ be strongly $\mathcal{O}$-convex. We show that P is strongly $\widetilde{\mathcal{O}}$-convex, which implies that it is strongly $\mathcal{O}$-convex by Theorem 6.10. Thus, we have to demonstrate that, for every two points p and q of P, their $\widetilde{\mathcal{O}}$-block is in P. Since $\mathcal{H}$ is strongly $\mathcal{O}$-convex, it is an $\widetilde{\mathcal{O}}$-hyperplane; hence, $\widetilde{\mathcal{O}}$-block(p, q) is a subset of the $\mathcal{H}$-layer of p and q, which is wholly in P.

Now suppose, conversely, that the boundary $\mathcal{H}$ of a halfspace P is not strongly $\mathcal{O}$-convex. Then, there are points p and q in $\mathcal{H}$ such that $\mathcal{O}$-block(p, q) is not in $\mathcal{H}$. Let x be the midpoint of the line segment joining p and q, as shown in Fig. 6.9. Furthermore, let y be a point of $\mathcal{O}$-block(p, q) outside of $\mathcal{H}$, and

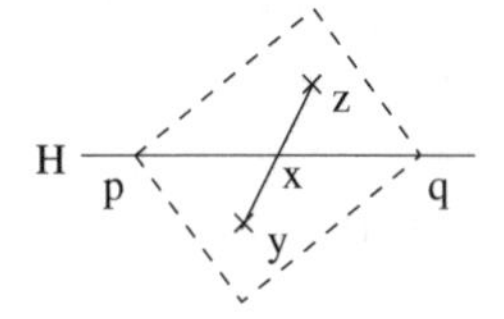

Fig. 6.9. Proof of Theorem 6.14

let z be the point on the line through x and y such that (1) y and z are on different sides of x and (2) the distance from z to x is the same as the distance from y to x.

Since y belongs to every $\mathcal{O}$-layer of p and q, we readily conclude that z also belongs to every $\mathcal{O}$-layer of p and q, which implies that z is in $\mathcal{O}$-block(p, q). We next note that x is in $\mathcal{H}$, which implies that y and z are on different sides of $\mathcal{H}$; hence, either y or z is outside of P. Thus, the points p and q are in P, whereas $\mathcal{O}$-block(p, q) is not completely in P, which implies that P is not strongly $\mathcal{O}$-convex. □

We next consider supporting hyperplanes of strongly $\mathcal{O}$-convex sets. A hyperplane **supports** a set if it "touches" the set at some boundary points and does not partition the set into two regions. For example, if we place a three-dimensional object on a table, then the table surface supports the object. Formally, a hyperplane $\mathcal{H}$ supports a set P if (1) the intersection of $\mathcal{H}$ with the boundary of P is nonempty and (2) P is contained in one of the two halfspaces whose boundary is $\mathcal{H}$.

Standard convex sets can be characterized in terms of supporting hyperplanes; specifically, *a closed set with a nonempty interior is convex if and only if, for every point of its boundary, there is a supporting hyperplane through it.* We generalize this property to strong $\mathcal{O}$-convexity.

Theorem 6.15. *A closed set with a nonempty interior is strongly $\mathcal{O}$-convex if and only if, for every point of its boundary, there is a strongly $\mathcal{O}$-convex hyperplane through it that supports the set.*

Proof. Let P be a closed set with a nonempty interior. Suppose that, for every point x of P's boundary, there is a strongly $\mathcal{O}$-convex hyperplane through x that supports the set. Then, for every boundary point x, there is a strongly $\mathcal{O}$-convex halfspace with boundary through x that contains P. The intersection of all such halfspaces is the set P, which implies that P is strongly $\mathcal{O}$-convex.

Suppose, conversely, that P is strongly $\mathcal{O}$-convex and let x be a point in its boundary. Since P is convex, its boundary in some neighborhood of x can be viewed as the graph of some convex function f.

First, suppose that the function f is differentiable at the point x. Then, there is exactly one supporting hyperplane $\mathcal{H}$ through x, and we need to prove

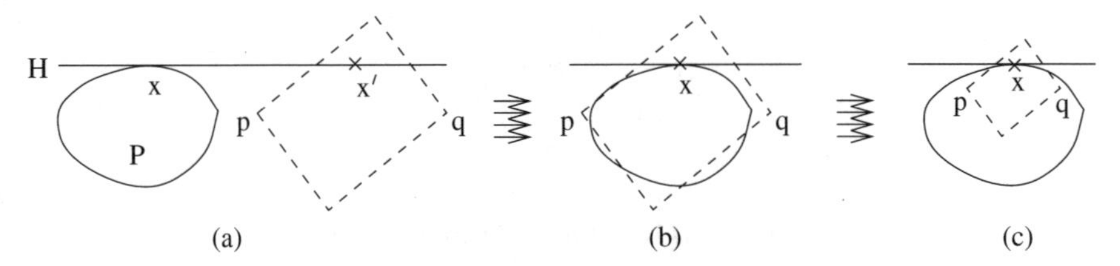

Fig. 6.10. Proof of Theorem 6.15

that it is strongly $\mathcal{O}$-convex. For convenience, assume that $\mathcal{H}$ is horizontal and P is below $\mathcal{H}$, as shown in Fig. 6.10a.

If $\mathcal{H}$ is not strongly $\mathcal{O}$-convex, then the halfspace with boundary $\mathcal{H}$ that contains P is not strongly $\mathcal{O}$-convex by Theorem 6.14. Therefore, there are points p and q in this halfspace such that $\mathcal{O}$-block(p, q) is not in the halfspace, as shown in Fig. 6.10a. Without loss of generality, assume that p and q are not in $\mathcal{H}$; if p or q is in $\mathcal{H}$, then we can move these points down a little in such a way that part of $\mathcal{O}$-block(p, q) remains above $\mathcal{H}$.

We choose some point $x' \in \mathcal{O}$-block$(p, q) \cap \mathcal{H}$ and translate $\mathcal{O}$-block(p, q) in such a way that x' becomes identical to x, as shown in Fig. 6.10b. Next, we scale (that is, homothetically transform) $\mathcal{O}$-block(p, q) in such a way that the point x' of the $\mathcal{O}$-block remains identical to x, as shown in Fig. 6.10c. Since the function f is differentiable at x, for a sufficiently small scaled version of the $\mathcal{O}$-block, the points p and q are below the graph of the function; that is, they are in P, as shown in Fig. 6.10c. On the other hand, part of the scaled version of $\mathcal{O}$-block(p, q) is above $\mathcal{H}$ and, hence, outside of P. Since a translate and a scaled version of an $\mathcal{O}$-block is an $\mathcal{O}$-block, we conclude that there are two points in P such that their $\mathcal{O}$-block is not in P, which contradicts the assumption that P is strongly $\mathcal{O}$-convex.

Next, suppose that the function f is not differentiable at the point x. Then, there may be more than one supporting hyperplane through x. We need to show that at least one of these hyperplanes is strongly $\mathcal{O}$-convex.

Since f is a convex function, it is a function of locally bounded variation, which implies that, in some neighborhood of x, the set of points where f is not differentiable is of measure zero. Therefore, there is a sequence of points where f is differentiable that is convergent to x. The supporting hyperplane through each of these points is strongly $\mathcal{O}$-convex.

We select a convergent subsequence from this sequence of supporting hyperplanes, and let $\mathcal{H}$ be the limit of the subsequence. Then, $x \in \mathcal{H}$ and, since the set $\widetilde{\mathcal{O}}$ of strongly $\mathcal{O}$-convex hyperplanes is closed by Corollary 6.11, $\mathcal{H}$ is strongly $\mathcal{O}$-convex. We show, by contradiction, that $\mathcal{H}$ supports P. If $\mathcal{H}$ does not support P, then, since P is convex, $\mathcal{H}$ intersects the interior of P. Let y be an interior point of P that belongs to $\mathcal{H}$, and $B_x \subseteq P$ be an open ball centered

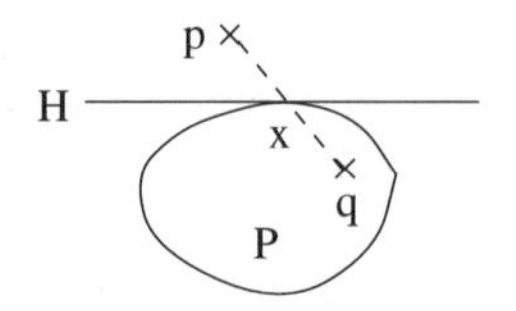

Fig. 6.11. Proof of Theorem 6.16

at x. There is a hyperplane in the convergent subsequence that intersects B_x; this hyperplane does not support P, yielding a contradiction. □

To see that the analogous result does not hold for sets with empty interiors, consider an $\mathcal{O}$-plane H in three dimensions and a nonconvex set P contained in H. For every point in P's boundary, H is a supporting plane through this point; however, P is not strongly $\mathcal{O}$-convex.

We next characterize strongly $\mathcal{O}$-convex sets with nonempty interiors in terms of the intersections of strongly $\mathcal{O}$-convex halfspaces.

Theorem 6.16. *A closed set with a nonempty interior is strongly $\mathcal{O}$-convex if and only if it is either the entire space $\mathcal{R}^d$ or the intersection of strongly $\mathcal{O}$-convex halfspaces.*

Proof. Clearly, the intersection of strongly $\mathcal{O}$-convex halfspaces is strongly $\mathcal{O}$-convex. Now suppose that P is a strongly $\mathcal{O}$-convex set with a nonempty interior, and that it is not the entire space $\mathcal{R}^d$. To demonstrate that P is the intersection of strongly $\mathcal{O}$-convex halfspaces, we show that, for every point p outside of P, there is a strongly $\mathcal{O}$-convex halfspace that contains P and does not contain p.

Let p be a point outside of P, let q be an interior point of P, and let x be a point of the intersection of the line segment joining p and q with P's boundary, as shown in Fig. 6.11; since P is closed, $x \neq p$. By Theorem 6.15, there is a strongly $\mathcal{O}$-convex hyperplane $\mathcal{H}$ through x that supports P, shown by the solid line in Fig. 6.11. Since q is an interior point of P, we conclude that $q \notin \mathcal{H}$.

We next show that $p \notin \mathcal{H}$. Note that x is in $\mathcal{H}$, and that the points p, x, and q are on a line. If p were in $\mathcal{H}$, then $\mathcal{H}$ would contain the line through p and x, which implies that q would also be in $\mathcal{H}$, yielding a contradiction. We conclude that P and p are on different sides of $\mathcal{H}$, which means that the halfspace with boundary $\mathcal{H}$ that contains P does not contain p. □

We can readily generalize Theorem 6.16 to nonclosed sets if we consider **open halfspaces,** that is, halfspaces that do not contain their boundaries. *A set with a nonempty interior is strongly $\mathcal{O}$-convex if and only if it is either the entire space $\mathcal{R}^d$ or the intersection of strongly $\mathcal{O}$-convex open halfspaces.*

If $\mathcal{O}$ is finite, the boundaries of $\mathcal{O}$-convex halfspaces are $\mathcal{O}$-hyperplanes, and the intersection of such $\mathcal{O}$-halfspaces is a convex polytope whose facets are parallel to elements of $\mathcal{O}$. Thus, the following result describes strongly $\mathcal{O}$-convex objects for finite orientation sets.

Corollary 6.17. *For finite $\mathcal{O}$, a set with a nonempty interior is strongly $\mathcal{O}$-convex if and only if it is either the entire space $\mathcal{R}^d$ or a convex polytope whose facets are parallel to elements of $\mathcal{O}$.*

If $\mathcal{O}$ is an infinite orientation set, a polytope may be strongly $\mathcal{O}$-convex even if its facets are not parallel to elements of $\mathcal{O}$. For instance, if $\mathcal{O}$ is a (countable or uncountable) set whose closure comprises all hyperplanes through o, then strong $\mathcal{O}$-convexity is equivalent to standard convexity, which implies that every convex polytope is strongly $\mathcal{O}$-convex.

We now state a condition under which closed strongly $\mathcal{O}$-convex sets, even those with an empty interior, are formed by the intersections of strongly $\mathcal{O}$-convex halfspaces.

Theorem 6.18. *The following two conditions are equivalent; that is, if one of them holds, then the other also holds:*

1. *Every strongly $\mathcal{O}$-convex flat is the intersection of strongly $\mathcal{O}$-convex hyperplanes.*
2. *Every closed strongly $\mathcal{O}$-convex set is either the entire space $\mathcal{R}^d$ or the intersection of strongly $\mathcal{O}$-convex halfspaces.*

Proof. Suppose that every closed strongly $\mathcal{O}$-convex set is either the entire space $\mathcal{R}^d$ or the intersection of strongly $\mathcal{O}$-convex halfspaces, and consider a strongly $\mathcal{O}$-convex flat η. Note that, if a halfspace contains η, then η is either wholly in the interior of the halfspace or wholly in its boundary. We consider the collection C of all strongly $\mathcal{O}$-convex halfspaces whose boundaries contain η. Clearly, the intersection of these halfspaces is equal to the intersection of *all* strongly $\mathcal{O}$-convex halfspaces that contain η; this intersection is exactly η. Since η is contained in the boundary of every halfspace in C, we conclude that η is the intersection of the boundaries of the halfspaces in C. By Theorem 6.14, the boundaries of strongly $\mathcal{O}$-convex halfspaces are strongly $\mathcal{O}$-convex; hence, η is the intersection of strongly $\mathcal{O}$-convex hyperplanes.

Now suppose, conversely, that every strongly $\mathcal{O}$-convex flat is the intersection of strongly $\mathcal{O}$-convex hyperplanes, and consider a strongly $\mathcal{O}$-convex set P. If the interior of P is nonempty, then P is either the entire space $\mathcal{R}^d$ or the intersection of strongly $\mathcal{O}$-convex halfspaces by Theorem 6.16. If its interior is empty, then its affine hull is a proper subset of the space $\mathcal{R}^d$ by Proposition 6.12; in this case, let η be the affine hull of P, and k be the dimension of η. Since η is a lower-dimensional space, we can speak of halfspaces within it, called η-halfflats. We prove that P is the intersection of strongly $\mathcal{O}$-convex halfspaces in two steps. First, we show that P is either the entire flat η or the intersection of strongly $\mathcal{O}$-convex η-halfflats. Second, we show that the flat η is the intersection of strongly $\mathcal{O}$-convex halfspaces, and every strongly $\mathcal{O}$-convex η-halfflat is also the intersection of strongly $\mathcal{O}$-convex halfspaces.

We treat η as an independent k-dimensional space and define the orientation set $\mathcal{O}_\eta$ in this space, as described in Sect. 4.1; that is, a $(k-1)$-dimensional flat $H \subseteq \eta$ is an $\mathcal{O}_\eta$-flat if it is the intersection of η with some $\mathcal{O}$-hyperplane. Every $\mathcal{O}$-hyperplane that intersects and does not contain η gives rise to a $(k-1)$-dimensional $\mathcal{O}_\eta$-flat by Proposition 4.1. For every two points p and q of η, a set is an $\mathcal{O}_\eta$-layer of p and q if and only if it is the intersection of η with some $\mathcal{O}$-layer of p and q, which implies that $\mathcal{O}_\eta$-block$(p, q) = \eta \cap \mathcal{O}$-block$(p, q)$. Since η is strongly $\mathcal{O}$-convex by Lemma 6.13, $\mathcal{O}$-block(p, q) is in η, which implies that $\mathcal{O}_\eta$-block$(p, q) = \mathcal{O}$-block(p, q). We conclude that a set contained in η is strongly $\mathcal{O}_\eta$-convex if and only if it is strongly $\mathcal{O}$-convex; therefore, P is strongly $\mathcal{O}_\eta$-convex. The relative interior of P in η is nonempty by Proposition 6.12; hence, P is either the entire flat η or the intersection of strongly $\mathcal{O}$-convex η-halfflats by Theorem 6.16.

We next show that η is the intersection of strongly $\mathcal{O}$-convex halfspaces, and every strongly $\mathcal{O}$-convex η-halfflat is also the intersection of strongly $\mathcal{O}$-convex halfspaces. Since η is strongly $\mathcal{O}$-convex, it is the intersection of strongly $\mathcal{O}$-convex hyperplanes, and every strongly $\mathcal{O}$-convex hyperplane is the intersection of two strongly $\mathcal{O}$-convex halfspaces, which implies that η is the intersection of strongly $\mathcal{O}$-convex halfspaces. We now consider a strongly $\mathcal{O}$-convex η-halfflat Q, and let H be the boundary of Q in η; note that H is a $(k-1)$-dimensional flat. Since Q is strongly $\mathcal{O}_\eta$-convex, its boundary H is strongly $\mathcal{O}_\eta$-convex by Theorem 6.6, which implies that H is strongly $\mathcal{O}$-convex; therefore, H is the intersection of strongly $\mathcal{O}$-convex hyperplanes. At least one of these hyperplanes, say $\mathcal{H}$, does not contain η; the η-halfflat Q is the intersection of η and a halfspace with boundary $\mathcal{H}$. Since η is the intersection of strongly $\mathcal{O}$-convex halfspaces, we conclude that Q is also the intersection of strongly $\mathcal{O}$-convex halfspaces. □

We readily conclude from Theorem 6.8 that, in two and three dimensions, all strongly $\mathcal{O}$-convex flats are the intersections of strongly $\mathcal{O}$-convex hyperplanes. By Theorem 6.18, this observation implies that, *in two and three dimensions, every strongly $\mathcal{O}$-convex set is the intersection of the strongly $\mathcal{O}$-convex halfspaces that contain it.*

If $\mathcal{O}$ is a finite or closed countably infinite orientation set in higher dimensions, then every strongly $\mathcal{O}$-convex flat is an $\mathcal{O}$-flat by Theorem 6.7, which implies that every strongly $\mathcal{O}$-convex flat is the intersection of strongly $\mathcal{O}$-convex hyperplanes. Therefore, *for closed countable $\mathcal{O}$, all strongly $\mathcal{O}$-convex sets are also the intersections of strongly $\mathcal{O}$-convex halfspaces.*

Summary

We have extended strong $\mathcal{O}$-convexity to higher dimensions and explored its properties. In particular, we have shown that, for every point in the boundary of a strongly $\mathcal{O}$-convex set, there is a supporting strongly $\mathcal{O}$-convex hyperplane through it (Theorem 6.15). Moreover, every closed strongly $\mathcal{O}$-convex set with a nonempty interior is either the entire space $\mathcal{R}^d$ or the intersection of strongly $\mathcal{O}$-convex halfspaces (Theorem 6.16). In addition, we have characterized strongly $\mathcal{O}$-convex flats in terms of $\mathcal{O}$-flats (Theorem 6.6), established strong $\mathcal{O}$-convexity of the affine hull of a strongly $\mathcal{O}$-convex set (Lemma 6.13), and derived a condition for the equivalence of strong convexity induced by two different orientation sets (Theorem 6.10).

7

Closing Remarks

We have defined two generalizations of convexity in higher dimensions, called $\mathcal{O}$-convexity and strong $\mathcal{O}$-convexity, and investigated their properties. We conclude with a summary of the main results (Sect. 7.1), related conjectures (Sect. 7.2), and directions for future research (Sect. 7.3).

7.1 Main Results

The basic properties of $\mathcal{O}$-convexity, $\mathcal{O}$-connectedness and strong $\mathcal{O}$-convexity are similar to those of standard convexity; in Tables 7.1 and 7.2, we summarize the main results generalized from standard convexity.

Other important results on $\mathcal{O}$-convexity include the description of $\mathcal{O}$-convex and $\mathcal{O}$-connected sets through their intersections with $\mathcal{O}$-hyperplanes (Theorems 4.6 and 4.8), the segment-extension and cut-and-paste properties of $\mathcal{O}$-connected curves (Lemmas 4.12 and 4.13), and the characterization of disconnected $\mathcal{O}$-halfspaces in terms of their components (Theorem 5.2).

The main properties of strong $\mathcal{O}$-convexity not included in Tables 7.1 and 7.2 are the description of strongly $\mathcal{O}$-convex flats in terms of $\mathcal{O}$-flats (Theorem 6.6), the necessary and sufficient condition for the equivalence of orientation sets (Theorem 6.10), and the strong $\mathcal{O}$-convexity of the affine hull of a strongly $\mathcal{O}$-convex set (Lemma 6.13).

In addition, we have derived the conditions under which the two generalized convexities are equivalent to standard convexity (Lemmas 4.5 and 6.1, and Corollary 6.9). In particular, if the orientation set $\mathcal{O}$ is closed, the following three conditions are equivalent; that is, each condition implies the other two:

1. Every line is an $\mathcal{O}$-line.
2. Every $\mathcal{O}$-convex set is standard convex.
3. Every standard convex set is strongly $\mathcal{O}$-convex.

Table 7.1. Comparison of different convexities; also see Table 7.2

Intersection	
Standard convexity:	The intersection of convex sets is a convex set.
$\mathcal{O}$-convexity:	The intersection of $\mathcal{O}$-convex sets is an $\mathcal{O}$-convex set.
Strong $\mathcal{O}$-convexity:	The intersection of strongly $\mathcal{O}$-convex sets is a strongly $\mathcal{O}$-convex set.
Line intersection	
Standard convexity:	A set is convex if and only if its intersection with every line is connected.
$\mathcal{O}$-convexity:	By definition, a set is $\mathcal{O}$-convex if and only if its intersection with every $\mathcal{O}$-line is connected.
Strong $\mathcal{O}$-convexity:	A set is strongly $\mathcal{O}$-convex if and only if its intersection with every simple $\mathcal{O}$-connected curve is connected.
Visibility	
Standard convexity:	A set is convex if and only if, for every two of its points, the line segment joining them is wholly in the set.
$\mathcal{O}$-convexity:	A closed path-connected set is $\mathcal{O}$-convex if and only if every two of its points can be joined by a simple $\mathcal{O}$-convex segment that is wholly in the set.
$\mathcal{O}$-connectedness:	A closed set is $\mathcal{O}$-path connected if and only if every two of its points can be joined by a simple $\mathcal{O}$-connected segment that is wholly in the set.
Strong $\mathcal{O}$-convexity:	A set is strongly $\mathcal{O}$-convex if and only if every simple $\mathcal{O}$-connected segment joining its points is wholly in the set.
Supporting planes	
Standard convexity:	A closed set with a nonempty interior is convex if and only if, for every point of its boundary, there is a supporting hyperplane through this point.
Strong $\mathcal{O}$-convexity:	A closed set with a nonempty interior is strongly $\mathcal{O}$-convex if and only if, for every boundary point, there is a strongly $\mathcal{O}$-convex hyperplane through it that supports the set.

Note that, if the orientation set $\mathcal{O}$ satisfies these conditions, then generalized halfspaces are identical to standard halfspaces, and curvilinear segments in the visibility results are identical to line segments.

7.2 Conjectures

We first discuss the **contractability** of $\mathcal{O}$-connected sets and connected $\mathcal{O}$-halfspaces. Intuitively, a contractable set is connected and has no holes; for instance, flats and balls are contractable. On the other hand, a hollow sphere is not contractable because it has a cavity, and a doughnut (torus) is not

Table 7.2. Comparison of different convexities; also see Table 7.1

Halfspace convexity	
Standard convexity:	Every halfspace is convex.
$\mathcal{O}$-convexity:	Every $\mathcal{O}$-halfspace is $\mathcal{O}$-convex.
$\mathcal{O}$-connectedness:	If the orientation set $\mathcal{O}$ has the point-intersection property, then every directed $\mathcal{O}$-halfspace is $\mathcal{O}$-path connected.
Boundary convexity	
Standard convexity:	An interior-closed set is a halfspace if and only if its boundary is a nonempty convex set.
$\mathcal{O}$-convexity:	An interior-closed set with an $\mathcal{O}$-convex boundary is an $\mathcal{O}$-halfspace.
$\mathcal{O}$-connectedness:	If the orientation set $\mathcal{O}$ has the point-intersection property, then the boundary of a directed $\mathcal{O}$-halfspace is $\mathcal{O}$-connected.
Strong $\mathcal{O}$-convexity:	An interior-closed set is a strongly $\mathcal{O}$-convex halfspace if and only if its boundary is a nonempty strongly $\mathcal{O}$-convex set.
Complementation	
Standard convexity:	The closed complement of a halfspace is a halfspace.
$\mathcal{O}$-convexity:	The closed complement of an $\mathcal{O}$-halfspace is an $\mathcal{O}$-halfspace if and only if the boundary of the $\mathcal{O}$-halfspace is $\mathcal{O}$-convex.
$\mathcal{O}$-connectedness:	The closed complement of a directed $\mathcal{O}$-halfspace is a directed $\mathcal{O}$-halfspace.
Strong $\mathcal{O}$-convexity:	The closed complement of a strongly $\mathcal{O}$-convex halfspace is a strongly $\mathcal{O}$-convex halfspace.
Halfspace intersection	
Standard convexity:	A closed set is convex if and only if it is either the entire space or the intersection of halfspaces.
$\mathcal{O}$-connectedness:	We *conjecture* that every closed $\mathcal{O}$-connected set is the intersection of directed $\mathcal{O}$-halfspaces.
Strong $\mathcal{O}$-convexity:	A closed set with a nonempty interior is strongly $\mathcal{O}$-convex if and only if it is either the entire space or the intersection of strongly $\mathcal{O}$-convex halfspaces.

contractable because it has a hole. Formally, a set is contractable if it can be continuously transformed (contracted) into a point in such a way that all intermediate sets are contained in the original set, as shown in Fig. 7.1. For planar sets, contractability is equivalent to simple connectedness. In higher dimensions, contractability is a stronger property; for example, a hollow sphere is simply connected but not contractable.

Clearly, standard convex sets are contractable. Connected $\mathcal{O}$-convex sets in two dimensions are also contractable if the orientation set $\mathcal{O}$ is nonempty, as shown in Lemma 2.1. This property does not generalize to higher dimensions; for instance, the connected $\mathcal{O}$-convex set in Fig. 4.3d is not contractable.

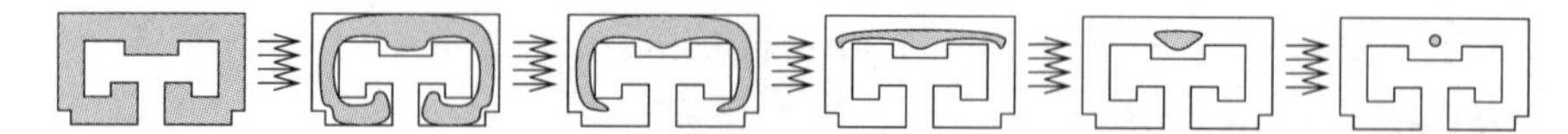

Fig. 7.1. Contraction of a set to a point

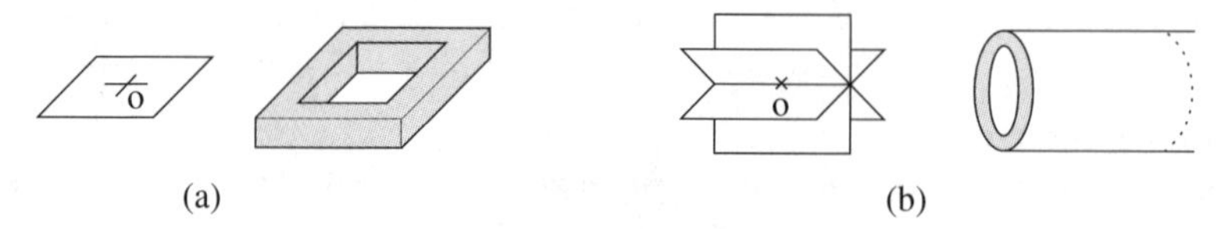

Fig. 7.2. $\mathcal{O}$-connected set **(a)** and $\mathcal{O}$-halfspace **(b)** that are not contractable

If the intersections of $\mathcal{O}$-hyperplanes do not form any $\mathcal{O}$-lines, then even $\mathcal{O}$-connected sets may not be contractable. For example, if the orientation set $\mathcal{O}$ in three dimensions includes only one plane, then the $\mathcal{O}$-connected set in Fig. 7.2a is not contractable. We conjecture that, if the orientation set $\mathcal{O}$ gives rise to $\mathcal{O}$-lines, then every $\mathcal{O}$-connected set is contractable.

The contractability of $\mathcal{O}$-halfspaces requires a stronger condition. If $\mathcal{O}$ does not have the point-intersection property, connected $\mathcal{O}$-halfspaces may not be contractable even when the set of $\mathcal{O}$-lines is nonempty. For example, the hollow "pipe" in Fig. 7.2b is an $\mathcal{O}$-halfspace that is not contractable. Establishing the contractability of $\mathcal{O}$-halfspaces for an orientation set with the point-intersection property is an open problem.

Conjecture 7.1 (Contractability).

1. *If the set of $\mathcal{O}$-lines is nonempty, every $\mathcal{O}$-connected set is contractable.*
2. *If the orientation set $\mathcal{O}$ has the point-intersection property, every connected $\mathcal{O}$-halfspace is contractable.*

The first part of this conjecture has a simple proof in three dimensions, illustrated in Fig. 7.3. Specifically, if a set is not contractable, then either its intersection with some $\mathcal{O}$-line is disconnected, as shown in Fig. 7.3a, or there is an $\mathcal{O}$-line l through a hole in the set, as shown in Fig. 7.3b. In the latter case, the intersection of the set with some $\mathcal{O}$-plane H containing l is disconnected, as shown in Fig. 7.3c. Thus, this set is not $\mathcal{O}$-connected in either case. Unfortunately, we have been unable to generalize this proof to higher dimensions.

Also, we have not generalized the halfspace-intersection property to $\mathcal{O}$-connectedness. Recall that every closed connected $\mathcal{O}$-convex set in two dimensions is the intersection of $\mathcal{O}$-halfplanes by Lemma 2.6, but this property does not generalize to $\mathcal{O}$-convex sets in higher dimensions, as shown by the counterexample in Fig. 5.4. We conjecture that $\mathcal{O}$-connected sets and connected $\mathcal{O}$-halfspaces are formed by the intersections of directed $\mathcal{O}$-halfspaces.

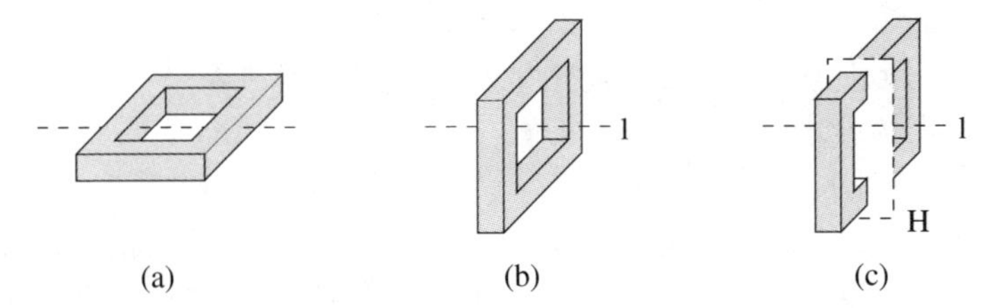

Fig. 7.3. Proof of the contractability conjecture in three dimensions

Conjecture 7.2 (Halfspace intersection).

1. *Every closed $\mathcal{O}$-connected set is the intersection of directed $\mathcal{O}$-halfspaces.*
2. *Every connected $\mathcal{O}$-halfspace is the intersection of directed $\mathcal{O}$-halfspaces.*

We next consider the boundaries of $\mathcal{O}$-convex polytopes. In two dimensions, if an orientation set contains n lines, the boundary of every $\mathcal{O}$-convex polygon can be partitioned into at most n $\mathcal{O}$-convex polygonal lines [35]. We conjecture that a similar result holds in higher dimensions; specifically, *for every finite orientation set $\mathcal{O}$, there is a fixed integer n such that the boundary of every $\mathcal{O}$-convex polytope can be partitioned into at most n connected $\mathcal{O}$-convex regions.*

The study of the boundaries of directed $\mathcal{O}$-halfspaces also suggests an interesting problem. Recall that, for orientation sets with the point-intersection property, the boundaries of directed $\mathcal{O}$-halfspaces are $\mathcal{O}$-connected by Theorem 5.14. The unsolved problem is *whether their boundaries are $\mathcal{O}$-path connected,* which would be a stronger property.

Next, note that we have established conditions of orientation-set equivalence for $\mathcal{O}$-convexity and strong $\mathcal{O}$-convexity, but we have not derived analogous results for $\mathcal{O}$-connectedness and $\mathcal{O}$-halfspaces.

Conjecture 7.3. *If $\mathcal{O}_2$ is the closure of $\mathcal{O}_1$, then a set is a directed $\mathcal{O}_1$-halfspace if and only if it is a directed $\mathcal{O}_2$-halfspace.*

If Conjectures 7.2 and 7.3 are both correct, then Conjecture 7.3 can be readily generalized to closed $\mathcal{O}$-connected sets and $\mathcal{O}$-halfspaces; that is, if $\mathcal{O}_2$ is the closure of $\mathcal{O}_1$, then every $\mathcal{O}_1$-connected set is $\mathcal{O}_2$-connected and every $\mathcal{O}_1$-halfspace is an $\mathcal{O}_2$-halfspace.

Finally, observe that, if two closed convex sets do not intersect, then there is a halfspace that contains one of them and does not intersect the other. We conjecture that an analogous result holds for $\mathcal{O}$-path connectedness and strong $\mathcal{O}$-convexity.

Conjecture 7.4 (Separation).

1. *If two closed $\mathcal{O}$-path-connected sets do not intersect, there is a directed $\mathcal{O}$-halfspace that contains one of them and does not intersect the other.*

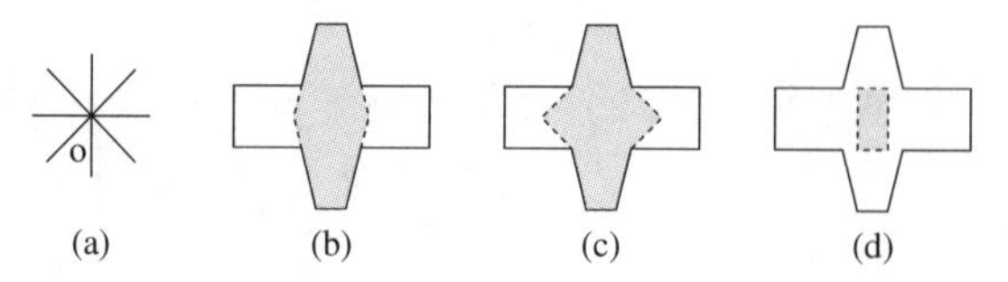

Fig. 7.4. Kernels (shaded regions) in standard convexity **(b)**, $\mathcal{O}$-convexity **(c)**, and strong $\mathcal{O}$-convexity **(d)**

2. *If two closed strongly $\mathcal{O}$-convex sets do not intersect, there is a strongly $\mathcal{O}$-convex halfspace that contains one of them and does not intersect the other.*

7.3 Future Work

In conclusion, we point out several unexplored areas of restricted-orientation convexity. First, note that each generalization of convexity gives rise to non-traditional kernels. Recall that the standard kernel of a geometric object is the set of points that are standardly visible from every point of the object. We can define an $\mathcal{O}$-convexity analog of visibility through $\mathcal{O}$-convex or $\mathcal{O}$-connected curves, and strong $\mathcal{O}$-visibility through $\mathcal{O}$-blocks, which lead to several types of kernels [35, 45]. In Fig. 7.4, we illustrate kernels for standard convexity, $\mathcal{O}$-convexity and strong $\mathcal{O}$-convexity.

Rawlins studied properties of kernels in planar $\mathcal{O}$-convexity and showed that, if $\mathcal{O} = \mathcal{O}_1 \cup \mathcal{O}_2$, then, for every contractable set P, $\mathcal{O}$-kernel(P) = $\mathcal{O}_1$-kernel$(P) \cap \mathcal{O}_2$-kernel(P) [35]; however, this result does not hold in higher dimensions. Later, Schuierer and Wood investigated $\mathcal{O}$-kernels of noncontractable sets in two dimensions [43, 46, 47]. A related open problem is to study the $\mathcal{O}$-kernels and strong $\mathcal{O}$-kernels of higher-dimensional sets.

Also, we have not studied $\mathcal{O}$-convex and $\mathcal{O}$-connected surfaces, whose role may prove analogous to that of hyperplanes in standard convexity. The study of these surfaces may be related to characterizing $\mathcal{O}$-convex sets in terms of their boundaries.

We may also explore alternative definitions of restricted-orientation convexity; for example, we may consider sets formed by the intersections of directed $\mathcal{O}$-halfspaces [26], characterize them in terms of their connected components, and investigate their relationship to $\mathcal{O}$-convex and $\mathcal{O}$-connected sets.

More general research directions include the extension of other generalized convexities, such as NESW convexity [35, 40] and link convexity [2, 52], to higher dimensions, as well as the exploration of computational aspects of restricted-orientation convexity, such as verifying the $\mathcal{O}$-convexity of given polytopes, and computing their $\mathcal{O}$-hulls and $\mathcal{O}$-kernels.

Churchill once said, “It is not the end, it is not even the beginning of the end, but it is the end of the beginning.” We believe that we have reached the end of the beginning with respect to restricted-orientation convexity.

References

1. Marshall W. Bern. Hidden surface removal for rectangles. *Journal of Computer and System Sciences*, 40(1):49–69, 1990.
2. C.K. Bruckner and J.B. Bruckner. On L_n-sets, the Hausdorff metric, and connectedness. *Proceedings of the American Mathematical Society*, 13(5):765–767, 1962.
3. Mark T. de Berg. On rectilinear link distance. *Computational Geometry: Theory and Applications*, 1:13–34, 1991.
4. Mark T. de Berg, Marc J. van Kreveld, and Jack Snoeyink. Two- and three-dimensional point location in rectangular subdivisions. *Journal of Algorithms*, 18(2):256–277, 1995.
5. Herbert Edelsbrunner. *Algorithms in Combinatorial Geometry*. Springer-Verlag, Berlin, Germany, 1987.
6. Moritz Werner Fenchel. Convexity through the ages. In Peter M. Gruber and Jörg M. Wills, editors, *Convexity and Its Applications*, pages 120–130. Birkhäuser Verlag, Boston, MA, 1988.
7. Eugene Fink and Derick Wood. Fundamentals of restricted-orientation convexity. *Information Sciences*, 92:175–196, 1996.
8. Eugene Fink and Derick Wood. Generalized halfspaces in restricted-orientation convexity. *Journal of Geometry*, 62:99–120, 1998.
9. Eugene Fink and Derick Wood. Strong restricted-orientation convexity. *Geometriae Dedicata*, 69(1):35–51, 1998.
10. Eugene Fink and Derick Wood. Planar strong visibility. *International Journal of Computational Geometry and Applications*, 13(2):173–187, 2003.
11. Michael T. Goodrich, Mikhail J. Atallah, and Mark H. Overmars. Output-sensitive methods for rectilinear hidden surface removal. *Information and Computation*, 107(1):1–24, 1993.
12. Ronald L. Graham. An efficient algorithm for determining the convex hull of a finite planar set. *Information Processing Letters*, 1(4):132–133, 1972.
13. Peter Gritzmann and Victor Klee. Computational convexity. In Jacob E. Goodman and Joseph O'Rourke, editors, *Handbook of Discrete and Computational Geometry*, pages 491–515. CRC Press, Boca Raton, FL, 1997.

14. Peter M. Gruber. History of convexity. In Peter M. Gruber and Jörg M. Wills, editors, *Handbook of Convex Geometry*, volume A, pages 1–15. Elsevier Science Publishers, Amsterdam, Netherlands, 1993.
15. Branko Grünbaum, Victor Klee, Micha A. Perles, and G. C. Shephard. *Convex Polytopes.* John Wiley and Sons, New York, NY, 1967.
16. Ralf Hartmut Güting. *Conquering Contours: Efficient Algorithms for Computational Geometry.* PhD thesis, Fachbereich Informatik, Universität Dortmund, 1983.
17. Ralf Hartmut Güting. Stabbing C-oriented polygons. *Information Processing Letters*, 16(1):35–40, 1983.
18. Ralf Hartmut Güting. Dynamic C-oriented polygonal intersections searching. *Information and Control*, 63(3):143–163, 1984.
19. Hugo Hadwiger, Hans Debrunner, and Victor Klee. *Combinatorial Geometry in the Plane.* Holt, Rinehart and Winston, New York, NY, 1964.
20. Paul Joseph Kelly and Max L. Weiss. *Geometry and Convexity: A Study in Mathematical Methods.* Krieger Publishing, Melbourne, FL, 1978.
21. Victor Klee. What is a convex set? *American Mathematical Monthly*, 78:616–631, 1971.
22. Der-Tsai Lee and I.M. Chen. Display of visible edges of a set of convex polygons. In Godfried T. Toussaint, editor, *Computational Geometry*, pages 249–265. Elsevier Science Publishers, Amsterdam, Netherlands, 1985.
23. Der-Tsai Lee and Franco P. Preparata. An optimal algorithm for finding the kernel of a polygon. *Journal of the ACM*, 26(3):415–421, 1979.
24. F.W. Levi. On Helly's theorem and the axioms of convexity. *Journal of the Indian Mathematical Society*, 15:65–76, 1951.
25. Witold Lipski and Christos H. Papadimitriou. A fast algorithm for testing for safety and detecting deadlocks in locked transaction systems. *Journal of Algorithms*, 2(3):211–226, 1981.
26. V. Martynchik, Nikolai N. Metelski, and Derick Wood. $\mathcal{O}$-convexity: Computing hulls, approximations, and orientation sets. In *Proceedings of the Eighth Canadian Conference on Computational Geometry*, pages 2–7, 1996.
27. Duncan McCallum and David Avis. A linear algorithm for finding the convex hull of a simple polygon. *Information Processing Letters*, 9(5):201–206, 1979.
28. Delfin Y. Montuno and Alain Fournier. Finding the x-y convex hull of a set of x-y polygons. Technical Report 148, Department of Computer Science, University of Toronto, 1982.
29. J. Ian Munro, Mark H. Overmars, and Derick Wood. Variations on visibility. In *Proceedings of the Third Annual Symposium on Computational Geometry*, pages 291–299, 1987.
30. Tina M. Nicholl, Der-Tsai Lee, Y. Z. Liao, and Chak-Kuen Wong. On the X-Y convex hull of a set of X-Y polygons. *BIT*, 23(4):456–471, 1983.
31. Bengt Julio Nilsson, Thomas Ottmann, Sven Schuierer, and Christian Icking. Restricted orientation computational geometry. In Burkhard Monien and Thomas Ottmann, editors, *Data Structures and Efficient Algorithms: Final Report on the DFG Special Joint Initiative*, LNCS Volume 594, pages 148–185. Springer-Verlag, Berlin, Germany, 1992.

32. Thomas Ottmann, Eljas Soisalon-Soininen, and Derick Wood. On the definition and computation of rectilinear convex hulls. *Information Sciences*, 33:157–171, 1984.
33. Mark H. Overmars and Chee-Keng Yap. New upper bounds in Klee's measure problem. *SIAM Journal on Computing*, 20(6):1034–1045, 1991.
34. Franco P. Preparata and Michael Ian Shamos. *Computational Geometry: An Introduction.* Springer-Verlag, Berlin, Germany, 1985.
35. Gregory J. E. Rawlins. *Explorations in Restricted-Orientation Geometry.* PhD thesis, School of Computer Science, University of Waterloo, 1987. Technical Report CS-87-57.
36. Gregory J. E. Rawlins, Sven Schuierer, and Derick Wood. Towards a general theory of visibility. In *Proceedings of the Second Canadian Conference on Computational Geometry*, pages 354–357, 1990.
37. Gregory J. E. Rawlins and Derick Wood. Optimal computation of finitely oriented convex hulls. *Information and Computation*, 72(2):150–166, 1987.
38. Gregory J. E. Rawlins and Derick Wood. Computational geometry with restricted orientations. In *Proceedings of the Thirteenth IFIP Conference on System Modeling and Optimization*, pages 375–384, 1988.
39. Gregory J. E. Rawlins and Derick Wood. Ortho-convexity and its generalizations. In Godfried T. Toussaint, editor, *Computational Morphology: A Computational Geometric Approach to the Analysis of Form*, pages 137–152. Elsevier Science Publishers, Amsterdam, Netherlands, 1988.
40. Gregory J. E. Rawlins and Derick Wood. A decomposition theorem for convexity spaces. *Journal of Geometry*, 36:143–159, 1989.
41. Gregory J. E. Rawlins and Derick Wood. Restricted-oriented convex sets. *Information Sciences*, 54(3):263–281, 1991.
42. Jörg-Rüdiger Sack. *Rectilinear Computational Geometry.* PhD thesis, School of Computer Science, McGill University, 1984.
43. Sven Schuierer. *On Generalized Visibility.* PhD thesis, Institut für Informatik, Universität Freiburg, 1991.
44. Sven Schuierer, Gregory J. E. Rawlins, and Derick Wood. Visibility, skulls, and kernels in convexity spaces. Technical Report CS-89-48, School of Computer Science, University of Waterloo, 1989.
45. Sven Schuierer, Gregory J. E. Rawlins, and Derick Wood. A generalization of staircase visibility. In *Proceedings of the International Workshop on Computational Geometry*, pages 277–287, 1991.
46. Sven Schuierer and Derick Wood. Restricted-orientation visibility. Technical Report 40, Institut für Informatik, Universität Freiburg, 1991.
47. Sven Schuierer and Derick Wood. Staircase visibility and computation of kernels. *Algorithmica*, 14(1):1–26, 1995.
48. Sven Schuierer and Derick Wood. Visibility in semi-convex spaces. *Journal of Geometry*, 60:160–187, 1997.
49. Eljas Soisalon-Soininen and Derick Wood. An optimal algorithm for testing for safety and detecting deadlocks in locked transaction systems. In *Proceedings of the ACM Symposium on Principles of Database Systems*, pages 108–116, 1982.
50. Eljas Soisalon-Soininen and Derick Wood. Optimal algorithms to compute the closure of a set of iso-rectangles. *Journal of Algorithms*, 5(2):199–214, 1984.

51. Frederick Albert Valentine. *Convex Sets.* McGraw-Hill, New York, NY, 1964.
52. Frederick Albert Valentine. Local convexity and L_n-sets. *Proceedings of the American Mathematical Society*, 16(6):1305–1310, 1965.
53. M.L.J. van de Vel. *Theory of Convex Structures.* Elsevier Science Publishers, Amsterdam, Netherlands, 1993.
54. Roger J. Webster. *Convexity.* Oxford University Press, New York, NY, 1994.
55. Peter Widmayer, Ying-Fung Wu, Martine D. F. Schlag, and Chak-Kuen Wong. On some union and intersection problems for polygons with fixed orientations. *Computing*, 36:183–197, 1986.
56. Peter Widmayer, Ying-Fung Wu, and Chak-Kuen Wong. Distance problems in computational geometry for fixed orientations. In *Proceedings of the First Annual ACM Symposium on Computational Geometry*, pages 186–195, 1985.
57. Peter Widmayer, Ying-Fung Wu, and Chak-Kuen Wong. On some distance problems in fixed orientations. *SIAM Journal on Computing*, 16(4):728–746, 1987.
58. Derick Wood. An isothetic view of computational geometry. In Godfried T. Toussaint, editor, *Computational Geometry*, pages 429–459. Elsevier Science Publishers, Amsterdam, Netherlands, 1985.
59. Derick Wood. The riches of rectangles. In *Proceedings of the Fifth International Meeting of Young Computer Scientists*, pages 161–168. Springer-Verlag, Berlin, Germany, 1988.
60. Derick Wood, Gregory J. E. Rawlins, and Sven Schuierer. Convexity, visibility, and orthogonal polygons. In Robert A. Melter, Azriel Rosenfeld, and Prabir Bhattacharya, editors, *Vision Geometry*, pages 225–237. American Mathematical Society, Providence, RI, 1991.

Druck:	Strauss Offsetdruck, Mörlenbach
Verarbeitung:	Schäffer, Grünstadt

Monographs in Theoretical Computer Science · An EATCS Series

K. Jensen
Coloured Petri Nets
***Basic Concepts,* Analysis Methods and Practical Use, Vol. 1**
2nd ed.

K. Jensen
Coloured Petri Nets
Basic Concepts, *Analysis Methods* and Practical Use, Vol. 2

K. Jensen
Coloured Petri Nets
Basic Concepts, Analysis Methods and *Practical Use,* Vol. 3

A. Nait Abdallah
The Logic of Partial Information

Z. Fülöp, H. Vogler
Syntax-Directed Semantics
Formal Models Based on Tree Transducers

A. de Luca, S. Varricchio
Finiteness and Regularity in Semigroups and Formal Languages

E. Best, R. Devillers, M. Koutny
Petri Net Algebra

S.P. Demri, E. S. Orłowska
Incomplete Information: Structure, Inference, Complexity

J.C.M. Baeten, C.A. Middelburg
Process Algebra with Timing

L.A. Hemaspaandra, L.Torenvliet
Theory of Semi-Feasible Algorithms

E. Fink, D. Wood
Restricted-Orientation Convexity

M. Hansen, C. Zhou
Duration Calculus

M. Große-Rhode
Semantic Integration of Heterogeneous Software Specifications

Texts in Theoretical Computer Science · An EATCS Series

J. L. Balcázar, J. Díaz, J. Gabarró
Structural Complexity I
2nd ed. (see also overleaf, Vol. 22)

M. Garzon
Models of Massive Parallelism
Analysis of Cellular Automata and Neural Networks

J. Hromkovič
Communication Complexity and Parallel Computing

A. Leitsch
The Resolution Calculus

G. Păun, G. Rozenberg, A. Salomaa
DNA Computing
New Computing Paradigms

A. Salomaa
Public-Key Cryptography
2nd ed.

K. Sikkel
Parsing Schemata
A Framework for Specification and Analysis of Parsing Algorithms

H. Vollmer
Introduction to Circuit Complexity
A Uniform Approach

W. Fokkink
Introduction to Process Algebra

K. Weihrauch
Computable Analysis
An Introduction

J. Hromkovič
Algorithmics for Hard Problems
Introduction to Combinatorial Optimization, Randomization, Approximation, and Heuristics
2nd ed.

S. Jukna
Extremal Combinatorics
With Applications in Computer Science

P. Clote, E. Kranakis
Boolean Functions and Computation Models

L. A. Hemaspaandra, M. Ogihara
The Complexity Theory Companion

C.S. Calude
Information and Randomness.
An Algorithmic Perspective
2nd ed.

J. Hromkovič
Theoretical Computer Science
Introduction to Automata, Computability, Complexity, Algorithmics, Randomization, Communication and Cryptography

A. Schneider
Verification of Reactive Systems
Formal Methods and Algorithms

Former volumes appeared as EATCS Monographs on Theoretical Computer Science

Vol. 5: W. Kuich, A. Salomaa
Semirings, Automata, Languages

Vol. 6: H. Ehrig, B. Mahr
Fundamentals of Algebraic Specification 1
Equations and Initial Semantics

Vol. 7: F. Gécseg
Products of Automata

Vol. 8: F. Kröger
Temporal Logic of Programs

Vol. 9: K. Weihrauch
Computability

Vol. 10: H. Edelsbrunner
Algorithms in Combinatorial Geometry

Vol. 12: J. Berstel, C. Reutenauer
Rational Series and Their Languages

Vol. 13: E. Best, C. Fernández C.
Nonsequential Processes
A Petri Net View

Vol. 14: M. Jantzen
Confluent String Rewriting

Vol. 15: S. Sippu, E. Soisalon-Soininen
Parsing Theory
Volume I: Languages and Parsing

Vol. 16: P. Padawitz
Computing in Horn Clause Theories

Vol. 17: J. Paredaens, P. DeBra, M. Gyssens, D. Van Gucht
The Structure of the Relational Database Model

Vol. 18: J. Dassow, G. Páun
Regulated Rewriting in Formal Language Theory

Vol. 19: M. Tofte
Compiler Generators
What they can do, what they might do, and what they will probably never do

Vol. 20: S. Sippu, E. Soisalon-Soininen
Parsing Theory
Volume II: LR(k) and LL(k) Parsing

Vol. 21: H. Ehrig, B. Mahr
Fundamentals of Algebraic Specification 2
Module Specifications and Constraints

Vol. 22: J. L. Balcázar, J. Díaz, J. Gabarró
Structural Complexity II

Vol. 24: T. Gergely, L. Úry
First-Order Programming Theories

R. Janicki, P. E. Lauer
Specification and Analysis of Concurrent Systems
The COSY Approach

O. Watanabe (Ed.)
Kolmogorov Complexity and Computational Complexity

G. Schmidt, Th. Ströhlein
Relations and Graphs
Discrete Mathematics for Computer Scientists

S. L. Bloom, Z. Ésik
Iteration Theories
The Equational Logic of Iterative Processes